Undergraduate Texts in Physics

Sebastiano Vasi · Ulderico Wanderlingh ·
Giuseppe Mandaglio

Open Source Tools for Physics Data Analysis

An Introduction

 Springer

Sebastiano Vasi [iD]
MIFT Department
OPENFIS S.R.L.—SPIN OFF
ACCADEMICO, LABORATORIO A2AT3
University of Messina
Messina, Italy

Ulderico Wanderlingh [iD]
MIFT Department
OPENFIS S.R.L.—SPIN OFF
ACCADEMICO, LABORATORIO A2AT3
University of Messina
Messina, Italy

Giuseppe Mandaglio [iD]
MIFT Department
University of Messina, and INFN Catania
Messina, Italy

ISSN 2510-411X ISSN 2510-4128 (electronic)
Undergraduate Texts in Physics
ISBN 978-3-032-00723-0 ISBN 978-3-032-00721-6 (eBook)
https://doi.org/10.1007/978-3-032-00721-6

This Springer imprint is published by the registered company Springer Nature Switzerland AG
The registered company address is: Gewerbestrasse 11, 6330 Cham, Switzerland

If disposing of this product, please recycle the paper.

Preface

Physics is one of the most important branch of science dealing with the description of natural phenomena by means of the study of matter, motion, energy, and force. In science, technology, engineering, and mathematics (STEM) courses, a fundamental part of the training process consists in different laboratory activities, collecting data and interpreting them. In this frame, the analysis of data is complementary to experimental and/or theoretical activities and it could be really important to have solid bases of programming. The aim of the book changes according to the reader. For students in middle- and high-schools, this book provides the basic knowledge of Operative Systems (OSs) and rudiments of programming allowing the approach to all STEM subjects; for university-level students, it is possible to have a manual complete enough to start a scientific career related to the acquisition, interpretation, and processing of data: in fact, this book provides the basics of programming for all users (even those who have not had the opportunity to study these topics previously), and subsequently approaches data science with applications to science and physics through analysis frameworks and methodologies for acquiring and analyzing data from experiments. In this book, we focus our attention on open source tools for physics data analysis because they give the change to unlock the real computation potentiality that students can achieve by their programming skills.

The book is mainly divided in two parts. Part I (Chaps. 1–3) gives the rudiments of programming in C++ and Python languages. Part II (Chaps. 4–6), Root, a powerful and open-source data analysis framework, and the universe of the Python libraries will be introduced as a complete tool for data analysis, showing their application in data science. A detailed overview of the chapters of the book is now presented.

Chapter 1 provides a description of the most common OSs, giving the motivation in using Linux-like OSs in Science and offering the reader the minimal skills to make use of the powerful Linux shells.

In Chap. 2, we supply fundamentals of procedural programming: variables, arrays, control structures, loop interactions, flow charts, lists, dictionaries, tuples, and so on. Furthermore, this chapter presents C++ and Python (compiled and interpreted codes) as reference programming languages, allowing the reader to better understand the topics covered through various examples and practical exercises.

Chapter 3 shows the importance of Object Oriented Programming (OOP) by means of the use of classes, furnishes the main features of classes and an explanation of how to proceed to implement and use them, and presents the benefits of OOP in huge scientific projects and large collaborations.

We built Chap. 4 thinking about the potential that students could have using open-source libraries in data analysis. This Chapter describes some of the most used analysis frameworks in science (such as CERN-ROOT, Jupyter, and Anaconda), explaining how to install the libraries of these environments and how to use it. In addition, here we present some analysis tools used in the presence of huge data samples: histogram computation and algebra, data modeling, and procedures.

Chapters 5 and 6 contain a lot of information about data elaboration and visualization, and experimental data acquisition and analysis. Chapter 5 describes how to analyze and represent data by using Root-classes, Python packages and gnuplot (linux graphics utility), furnishing several details about how Python allows to make use of all the framework and libraries (e.g., PyROOT) and offering several practical problems that can be solved with different approaches. Finally, Chap. 6 explains how to design different experiments (in physics and in science, generally speaking) with the use of microcomputers and microcontrollers implementing codes able to acquire, elaborate, and visualize data, showing how to use analysis tools in combination with hardware and data acquisition processes. This is very important in all experiments where the real-time monitoring of physical quantities is necessary.

All program codes, exercises, and their solutions presented in this book are available at the "https://github.com/sn-code-inside/open-source-tools-for-physics-data-analysisopen-source-tools-for-physics-data-analysis" GitHub repository.

Messina, Italy
June 2025

Sebastiano Vasi
Ulderico Wanderlingh
Giuseppe Mandaglio

Competing Interests The authors have no competing interests to declare that are relevant to the content of this manuscript.

Contents

Acronyms

CD-ROM   Compact Disc-Read-Only Memory
CLI   Command Line Interface
CPU   Central Processing Unit
DVD   Digital Versatile Disc, Digital Video Disc
FHS   Filesystem Hierarchy Standard
FTP   File Transfer Protocol
GNU   GNU's Not Unix!
GUI   Graphical User Interface
HPC   High-Performance Computing
IDLE   Integrated Development and Learning Environment
OOP   Object-Oriented Programming
OS   Operative System
PDF   Portable Document Format
PID   Process IDentifier
RAM   Random Access Memory
ROM   Read-Only Memory
STEM   Science, Technology, Engineering and Mathematics
UNIX   UNiplexed Information Computing System
USB   Universal Serial Bus

List of Figures

List of Tables

Part I
Basic Knowledge

Operating systems (OS) are the backbone of any computer system, managing hardware, and software resources and providing essential services for computer programs. Among the various operating systems, Linux stands out due to its open-source nature, robustness, and flexibility. Linux is widely used in servers, desktops, and embedded systems. It offers a variety of distributions (distros) like Ubuntu, Fedora, and CentOS, each tailored for different use cases. Linux's command-line interface (CLI) is powerful, allowing users to perform complex tasks efficiently.

Open-source programming refers to the practice of making the source code of software freely available for anyone to use, modify, and distribute. Python and C++ are two prominent open-source programming languages.

Python is known for its simplicity and readability, making it an excellent choice for beginners. It has a vast standard library and supports multiple programming paradigms, including procedural, functional, and object-oriented programming. Python is widely used in web development, data analysis, artificial intelligence, and scientific computing.

C++, on the other hand, is a powerful language that provides fine-grained control over system resources. It is an extension of the C programming language and includes object-oriented features. C++ is commonly used in system/software development, game development, and applications requiring high performance, such as real-time simulations.

Object-oriented programming (OOP) is a programming paradigm that organizes code into objects, which are instances of classes. Each object contains data (attributes) and functions (methods) that operate on the data. OOP offers several advantages:

- Modularity: Code is organized into discrete objects, making it easier to manage and understand.
- Reusability: Classes and objects can be reused across different programs, reducing redundancy.
- Scalability: OOP makes it easier to manage and scale large codebases by breaking down complex problems into smaller, manageable objects.

- Encapsulation: Data and methods are encapsulated within objects, protecting the internal state and promoting data integrity.
- Inheritance: New classes can inherit attributes and methods from existing classes, promoting code reuse and reducing duplication.

In summary, understanding the basics of operating systems, especially Linux, along with open-source programming languages like Python and C++, and the principles of object-oriented programming, can provide a strong foundation for anyone who want to analyze data of physics' experiments.

In this part of the book, we will show the most important arguments related to these topics, making use of examples and exercises to help the reader.

Chapter 1
Operative Systems and Linux Shells

1.1 Introduction

An Operating System (OS) is essentially the ultimate application that controls computer functions after the bootstrap and eventually acts as an interface with humans behind the screen. It is itself a program that manages and controls most of the other programs running on the computer, providing the users with a graphical interface (which in the simplest case reduces to a bare terminal window) based on iconographic representation of computer contents and functions, to be selected and launched via classical mouse gestures. As a consequence on a single PC mass storage memory, several OSs can be present simultaneously but only one is executed. It is also possible, for the running OS, to run other OS in 'emulation mode'.

In general, a given OS is tailored around a given processor architecture or devices to optimize performance. Some of the most common OSs, across various devices, include Microsoft Windows and Apple OSX for personal computers, Android and iOS for mobile devices, and Linux for personal computers, servers, and a range of other applications.

In the following, a nonexhaustive list of what can be, on average, considered pros and cons of most common OSs (Table 1.1).

1.2 Motivations for Using Linux-Like OSs in Science

Although Windows and, to a lesser extent, macOS are currently the most popular OSs, many users working in science prefer to use Linux-like systems. Also, in big research centers and international large-scale research facilities, Linux OS is the most common choice for instrument control and networking infrastructure. This is also the case of High-Performance Computing (HPC) clusters, including large faculty research clusters, thanks to Linux's intrinsic multi-user nature and robust support

© The Author(s), under exclusive license to Springer Nature Switzerland AG 2026

S. Vasi et al., *Open Source Tools for Physics Data Analysis*, Undergraduate Texts in Physics, https://doi.org/10.1007/978-3-032-00721-6_1

Table 1.1 Pros and cons of the most common Operating systems

OS	Pros.	Cons.
Microsoft Windows	Windows provides a familiar and intuitive graphical user interface A vast array of software applications is compatible with Windows Many popular games are developed for Windows Windows supports a wide range of hardware configurations	Historically, Windows has been more susceptible to malware and security threats Some users find the forced updates and reboots inconvenient The full version of Windows often comes with a price tag
MacOS	MacOS is known for its elegant and user-friendly interface The Mac App Store offers a curated selection of high-quality applications macOS benefits from a robust security model, reducing the risk of malware Seamless integration with other Apple devices and services	Limited customization compared to some other operating systems Apple computers running macOS tend to be relatively expensive Limited hardware choices, as macOS is designed to run on Apple hardware
Android	Android is based on open-source software, fostering innovation and customization The Google Play Store offers a vast selection of applications Android is used on various devices, from smartphones to tablets Seamless integration with Google's ecosystem of services	Different devices may run different versions of Android, leading to fragmentation Being open source can make it more susceptible to specific security issues
iOS	iOS is known for its sleek and user-friendly interface The App Store is curated, resulting in a generally higher quality of applications OS is considered more secure due to its closed ecosystem and app vetting process Apple provides regular and timely updates for iOS devices	Limited customization options compared to Android Apple devices tend to be more expensive than their counterparts
Unix	Known for stability and long uptime without crashes Highly secure with robust user permissions and access controls Runs on a variety of hardware platforms with minimal modification Supports multiple users and tasks concurrently The modular approach allows customization Powerful and flexible command line interface	Can be complex to learn and use, especially for beginners Not all software is available or compatible with Unix Not as user-friendly as other OS Some versions can be expensive Finding and understanding documentation can be difficult May require specific hardware configurations
Linux	Linux is open source version of Unix with a free kernel, allowing users to modify and distribute their own versions Linux is known for its stability, often used as a server operating system Linux has a strong security track record Can run on older hardware and uses resources efficiently	Some commercial software and games are not developed for Linux For beginners, Linux can have a steeper learning curve Limited driver support for specific hardware

for parallel processing. Many scientists may have several reasons for appreciating Linux, which we have tried to summarize here quickly but comprehensively.

Linux operating systems are available on many different distributions (distros), allowing great customization to fit users' needs. The most well-known distributions are likely Ubuntu, Fedora, Debian, Knoppix, Arch, Red Hat, Gentoo, and Slackware. Moreover, several variants have been developed for each of these distros to suit specific requirements.

Linux operating systems are known for their reliability and robustness. This is witnessed by the fact that a large chunk of the Internet, corporate servers, and other critical infrastructure around the world, including big data centers and cloud services, are powered by Linux and are largely used in high-traffic and critical applications.

Although not all possible applications used by scientists working in the most diverse fields are available for Linux, it is actually possible to find many of them, or similar applications for the same scope, developed by teams of volunteers. In the case of scientific research, this is often the case. This is an added value when it comes to dealing with the format of the output files produced on different equipment. In fact, in many cases, the instrument data format is owned by the producer brand, and it often has to be read, manipulated, and stored by proprietary software. Conversely, those alternative applications, being developed by and for a wider user community, can deal with most of the standard file formats and can also read and occasionally write many of the manufacturers' file formats.

Many open-source applications running in Linux systems are no less than the corresponding commercial software, at least in processing capabilities. The only flaw that can be found concerns the aesthetics of the user interface, which may appear too minimalist and not very smooth. Conversely, it is often possible to add 'scripts' or 'macros', generated by users, to an application to extend its capabilities. These 'add-ons' are usually included in the application or can be grabbed from the web in the appropriate forums and discussion groups.

Nowadays, most, if not all, of a scientist's experience is based on the use of computers. This starts from data acquisition, since modern measurement equipment is computer-based. Then, collected data have a digital format, and their storage and retrieval are managed by using a computer. Moreover, data representation, manipulation, and analysis are all performed on computers by using specific software tools. Computers are so pervasive in scientific investigation, and software development proceeds at such a high rate that researchers who use these tools passively are often frustrated and unable to develop their own skills to the maximum.

Linux gently forces users to interact directly with the machine about running processes, system resources, permissions, file systems, and so on. This gives the user the chance to understand computers at a deeper level. At the same time, Linux promotes a problem-solving mindset. It drives users to troubleshoot issues, search for solutions, and read documentation—skills crucial in many fields beyond computer science.

Working with Linux platforms requires adaptability and provides a foundation of knowledge that can be used in various technical and scientific fields. This adaptability is essential in the ever-changing technology landscape, where different skills can open doors to new opportunities and work experiences.

We can safely say that the engine at the heart of Linux is open-source software, with the different distributions available exemplifying this principle. Open source promotes collaboration, innovation, and transparency, inviting users to explore, modify, and contribute to projects. This experience provides practical skills and mirrors professional coding/programming environments . Engaging in open-source projects allows users to work in real-world scenarios, improving their problem-solving and teamwork skills, especially for younger people approaching this path.

The very concept of Open Source is linked to its large and active community of users. This supporting network includes developers and enthusiasts who provide tutorials, forums, and guides. This community is a valuable resource for beginners, offering guidance and a wealth of information to help navigate and resolve any mishaps. The community aspect of Linux ensures that users are never isolated, fostering a continuous learning environment where knowledge is freely shared.

Another crucial characteristic of Linux is its lightweight nature; a minimal distros can even fit in 8 Mb supports. This makes it more efficient and faster since, for lightweight distros, it is possible to run exclusively on computer RAM. This lightweight nature becomes essential on older hardware, breathing new life into obsolete systems. At the same time, it is possible to install fully flagged distros that are capable of cloning the Graphical User Interface (GUI) of commercial Operating Systems

What also sets Linux apart is the level of control it offers over system operations. Unlike other operating systems that hide many features, Linux encourages users to dig deeper, understand, and manage their systems comprehensively. This transparency and control allow users to optimize their systems for better performance and security, tailoring the environment to their needs.

In the computing industry, Linux makes working with programming languages and tools easy. With Python often pre-installed and other languages such as Java, C++ easily accessible, Linux is an ideal programming platform. Developer-friendly Linux distributions simplify tasks such as installing programming languages, setting up databases, configuring servers, and managing dependencies.

Package Manager tools seamlessly manage installed software. They provide users with an easy way to install, remove, and update packages on their system. They also find the correct version for a given architecture and resolve library dependencies. Although there are several package managers, the three most commonly used are apt, yum, and pacman.

Apt   (Advanced Packaging Tool). Debian-based package managers are the default package managers for Debian-based Linux distributions, such as Ubuntu, Linux Mint, and Debian itself.

Yum   (Yellowdog Updater Modified). Red Hat-based package managers. It is the default package manager for Red Hat-based Linux distributions, such as CentOS, Fedora, and Red Hat Enterprise Linux.

Pacman   (Package Manager). Linux-based package managers are the default package managers for Arch Linux and its derivatives, such as Manjaro and EndeavourOS.

This efficient package management in Linux development has been a turning point, especially in environments that require frequent software updates and the integration of multiple tools.

Expertise in Linux is also highly valued in the technology industry. Linux powers servers, embedded systems, and supercomputers, making it an essential skill for system administration, software development, DevOps, and more careers. Understanding Linux can open numerous career doors, providing competitive skills in a job market that increasingly values technical versatility and open-source expertise.

Linux's open-source nature means it is a collaborative creation of tech enthusiasts around the world. It supports multitasking, such as performing data analysis while handling other tasks. Additionally, Linux is renowned for its resilience against viruses and malware, ensuring data security. This security feature is essential in environments where data integrity and confidentiality are critical, such as scientific research and enterprise systems.

A key aspect of using Linux for science is mastering the file system. Linux structures files and directories in a hierarchical tree format, with the root directory symbolized by a slash ("/"). Commands such as "cd" help navigate this structure, making finding and managing data easier. Understanding the file system is critical to organizing data, ensuring that data scientists can efficiently locate and manipulate data sets.

The Command Line Interface (CLI) in Linux is a powerful tool, offering a level of control that users appreciate. While GUIs provide ease of use, the CLI allows for fine-grained control, essential for tasks such as data manipulation, scripting, and server management. Mastering the CLI is like an extra edge, improving productivity and accuracy when managing complex operations.

1.2.1 Linux File System

The Linux file system is more than just a place where files reside; it is a structured environment that follows a strict logic, essential for maintaining system order and efficiency. Linux employs a hierarchical file system based on a parent-child relationship. This structure is known as the Linux Filesystem Hierarchy Standard (FHS). The FHS serves as a reference that outlines the conventions for organizing a UNIX system's file layout, and other UNIX variants, including Linux, also use it. The Linux Foundation maintains this standard. Within the Linux file system hierarchy is a top-level directory called the root directory, represented by the symbol /. All other directories and files are organized under this root directory, making them hierarchically subordinate.

Below is an overview of the most common directories and their functions:

/—Root Directory   The root directory, identified by a single slash ("/"), plays a crucial role as it represents the top level of the Linux file system hierarchy. This directory contains subdirectories and files that form the backbone of the operating system, and it is the reference point from which all other directories branch out. Its integrity and security are essential for the stability and security of the entire system.

/bin—Essential Command Binaries   The /bin directory is the repository for executable files essential for basic system operations, such as booting, recovery, and maintenance in single-user mode. These executable files include some of the most commonly used tools in Linux, crucial for file navigation and management, and are accessible to all users.

/boot—System Boot Loader Files   The /boot directory houses critical files required for the operating system's boot process. These include the Linux kernel image, essential for loading the system, and boot configuration files and scripts, such as GRUB, which serve as the computer's first point of contact during hardware initialization.

/dev—Device Files   The /dev directory is a unique aspect of Unix-like systems, where devices are treated as files. This directory contains special files representing physical devices (such as hard drives and USB) and virtual devices (such as sockets and pipes), allowing the system and users to interact with them through standard file operations.

/etc.—Host-specific Configuration Files   /etc. is the central library of the system's configuration files. Every service program and startup script draws from this directory for its configurations, ranging from user management to network services and system daemons. Changes in this directory can significantly impact the system's operation and security.

/home—User Home Directories   The /home directory is the virtual place where users can store their documents, application configurations, and personal data. It serves as a personal workspace for users, providing a clear and secure separation of data among different accounts.

/lib—Shared Library Modules   The /lib directory contains shared libraries necessary to operate essential programs located in /bin and /sbin. Libraries are codes that can be used by multiple programs simultaneously, saving space and resources, and enabling easier software updates and maintenance.

/media—Removable Media Files   /media acts as the mount point for removable devices, such as CD-ROMs, DVDs, and USB flash drives. The operating system usually creates subdirectories in /media automatically when a device is connected, allowing users easy access to the contents of removable devices.

/mnt—Temporary Mounted Filesystems   Filesystems of various types are temporarily mounted in /mnt, often used by system administrators for temporary manual mounts. It serves as a workspace for accessing non-permanent filesystems, such as those shared over a network or external disks in temporary use.

/opt—Add-on Application Packages   The /opt directory is intended to host additional software and applications outside the Linux distribution's standard package management system. This may include commercial software packages and third-party applications that do not integrate tightly into the standard file system hierarchy.

/proc—Automatically Generated Files   The /proc directory is a distinctive feature of Linux, providing a dynamic view of the running system. This directory does not physically exist on the disk. Still, it is dynamically generated by the kernel, offering a real-time window into the system and running processes, including parameters like CPU and memory usage and internal kernel configurations.

/root—Home Directory for root user   The /root directory is the personal space reserved for the root user, the top administrative account in the Linux system. This separation from /home ensures that critical administrator files are kept secure and separate from standard user data.

/run—Run Time Program Data   /run is designed to hold system information needed from system startup until the file system is fully loaded and is vital for the bootprocess and system execution. It includes volatile data such as the PIDs of running processes and lock files, which help manage and coordinate the startup and interaction between various services.

/sbin—System Binaries   The /sbin directory is similar to /bin but contains binaries and programs that the system administrator generally executes. These tools are essential for system maintenance, diagnostics, and repair, including commands for filesystem management, hardware control, and other system administration tasks.

/srv—Site-specific Data Served by the System   /srv contains data that is served by the system for various services, such as web server data, FTP, and other data transfer protocols. This directory is often used to store data exposed to clients over the network.

/sys—Virtual Directory for System   Like /proc, the /sys directory is a virtual view provided by the kernel that exposes information about the system's hardware through the sysfs file system, allowing administrators to obtain detailed information about the hardware and perform certain system-level configurations.

/tmp—Temporary Files   /tmp is the storage for temporary files used by applications and the operating system. These data are usually intended to be maintained only for a session and are often deleted upon system reboot, providing automatic cleanup and preventing the accumulation of unnecessary data.

/usr—Read-only User Data   /usr is one of the major subdirectories and serves as a repository for user applications, shared libraries, documentation, and source code. Designed to be read-only during normal system operation, /usr provides a fundamental collection of resources for users and programs.

/var—Variable Files   The /var directory is intended to contain files subject to frequent changes, such as log files that record system activities, print spool files, pending emails, and other dynamic data. These files often grow without predefined limits, requiring careful disk space management.

1.3 Power in a Shell

As mentioned before, Linux's shell is a text-based CLI that allows users to interact with the operating system by typing commands. It acts as an intermediary between the user and the operating system, interpreting the commands the user enters and executing the corresponding system functions. Although most work is performed using more practical GUIs on modern computers, which ultimately execute shell commands, it is still useful for Unix system users to be familiar with the shell. Unix system critically depends on the shell since all internal procedures are shell procedures.

1.3.1 Common Shells

Many shells are available, listed in the `/etc./shells` file, and are mostly executable files found in `/bin`. The shell assigned to the user at login is associated with their username in the `/etc./passwd` file. Users can change their default shell using the `chsh` command, or change it temporarily using, for example, `csh` command for C shell opening a process with the new shell interpreting commands. The main shells that are found on Unix system are:

- Bourne Shell (sh): One of the earliest and simplest Unix shells, written by Steve Bourne for UNIX System 7 in 1979.
- C Shell (csh): Introduced C-like syntax and scripting capabilities, developed at Berkeley by Bill Joy for the Berkeley version of UNIX around 1981
- Korn Shell (ksh): Combined features of the Bourne and C shells, adding more scripting capabilities, developed at AT&T by David Korn around 1986.
- Bourne Again Shell (bash):, similar to the Bourne and Korn shells, widely used in Linux systems and part of the GNU Project maintained by the FSF (Free Software Foundation). Written by Brian Fox and freely available since 1988.
- TENEX C Shell (tcsh): An enhanced version of the C shell that provides a more user-friendly and interactive experience than csh while maintaining compatibility with C shell scripts.
- Z Shell (zsh): Known for its extended customization options and interactive use features.

The key features of Unix shell are:

- **Command Execution**: The primary function of a shell is to execute commands. Users type commands, which are interpreted by the current shell and passed to the operating system to perform tasks like file manipulation, program execution, and system control.
- **Environment Control**: The shell provides an environment where users can set variables that influence the behavior of processes. For example, environment variables like PATH determine where the system looks for executable files.

- **File Management**: Users can navigate the file system, create, delete, and modify files and directories.
- **Process Management**: Shells allow users to monitor system processes and stop their execution. Commands such as `ps`, `kill`, `top`, and `bg`/`fg`.
- **I/O Redirection and piping**: Users can redirect input and output (I/O) to and from files, devices, and other commands. E.g. the output of a command can be redirected to a file, or the input of a command can come from a file. Redirecting the output of one command to be used as the input to another is called piping.
- **Scripting**: Shell commands can be gathered into a script, which means users can write a text file as a sequence of commands) to automate repetitive tasks. Inside a shell script, the user can include loops, conditionals, and variables, making them powerful tools for managing system operations.
- **Wildcard**: wildcards, also known as "globbing" characters, are special characters used in commands to represent or match a group of files or directories based on specific text structures. They are especially useful when dealing with multiple files simultaneously, allowing for more flexible and powerful command-line operations. Wildcard significantly improves efficiency when working in the terminal and deserves a detailed list:

`*` (Asterisk)   Matches zero or more characters.
Example: `ls *.txt` lists all files ending with `.txt`.

`?` (Question Mark)   Matches exactly one character.
Example: `ls file?.txt` matches `file1.txt` but not `file10.txt`.

`[ ]` (Square Brackets)   Matches any one of the characters inside the brackets.
Example: `ls file[123].txt` matches `file1.txt`, `file2.txt`, or `file3.txt`.

`[! ]` (Exclamation within Square Brackets)   Matches any character *not* inside the brackets.
Example: `ls file[!123].txt` matches `file4.txt`, `file5.txt`, etc., but not `file1.txt`, `file2.txt`, or `file3.txt`.

`{}` (Curly Braces)   Used for pattern expansion to match multiple alternatives.
Example: `mv file{1,2}.txt /backup/` moves `file1.txt` and `file2.txt` to `/backup/`.

`~` (Tilde)   Represents the home directory of the current user.
Example: `cd ~/Documents` changes the directory to the "Documents" folder in the user's home directory.

`**` (Double Asterisk in Bash 4.0+)   Matches files and directories recursively.
Example: `ls **/*.txt` lists all `.txt` files in the current directory and all subdirectories.

It is possible to combine wildcards to create more complex patterns.

'ls file[0-9]*.txt': Lists files that start with "file", followed by a digit, and end with '.txt'.

'cp /Documents/*.pdf /backup/': Copies all PDF files from the 'Documents' directory to the '/backup/' directory.

If you need to use a character that is normally a wildcard (such as * or ?) as a literal character, you can avoid expansion by preceding it with a backslash ('\'). 'ls file*.txt': Matches a file named 'file*.txt'.

All the shell commands have a common syntax:

```
command_name option option option ... argument argument
    ...
```

Options or flags are generally preceded by a - or -- and can sometimes be grouped together. For example, to list all the files that start with "a", ordered by creation date, the user will type:

```
ls -lt a*
```

In the following, we give lists of the most used shell commands; moreover, by pressing the Tab, after inputting the initial part of a command/ filename, the shell completes the command/filename or prompts the available commands/filenames that start with the typed text.

As has been told, a shell command runs an executable file (process) and locks the terminal windows till the end of the process; this can be avoided by typing command & to execute it in the background.

1.3.1.1   General Operations

Basic operations that can be performed in shells are:

exit   Exits the shell.
Ctrl+c   Interrupts a command.
Ctrl+a   Moves the cursor to the end of the line.
Ctrl+e   Moves the cursor to the end of the line.
Ctrl+d   Log you out of any terminal and close it.
;   Separates multiple commands given on the same line.
Indicates a comment after the command.
history   Displays the last executed commands (stored in the ~/.bash_history file). The hystory can also be accessed by pressing the up arrow key

1.3.1.2   Variable Operations

= Assigns a variable.
$ Evaluates a variable.

1.3.1.3 Obtaining Information

help   General information about the shell.
man command   Information about the command.
which command   Shows where the executable is.

1.3.1.4 Listing Files

ls a*   Shows files starting with 'a'.
tree -d   Shows directory tree.
stat filename   Shows all characteristics of a file or link.
lstat   or stat -1 Refers to the real file, not the link.
lsof   Shows all open files in the system.

Thanks to many options, Linux commands are highly versatile and offer numerous additional functionalities. These options can be activated by typing a space after the command, followed by a hyphen and one or more letters. It is recommended that the reader consult the manual of each command used by typing *man* followed by the name of the command they wish to learn more about.

1.3.1.5 Piping and Redirection

command >file   The standard output of the command goes into the file.
command »file   The command's standard output is appended to the file.
command <file   The command's standard input is the file.
command |command   Commands are concatenated; the standard output of the first
    is the standard input of the second.

In the example shown in Fig. 1.1, we can better understand how concatenated commands operate. A simplified explanation might state that the > operator creates the file list.dat and contains the ls command output. However, this description is incomplete and does not clarify why the file appears in the list. A more accurate explanation is that the > operator first creates the file list.dat, and only afterward is the ls command executed, whose output is redirected into the newly created file. Since the file exists at the time of execution of ls, its name is included in the output.

1.3.1.6 Navigating the File System

tree -d   Shows directory tree.shell!tree
cd directory   Changes to the directory.
cd ..   Goes to the parent directory.
cd ~   Goes to the home directory.

Fig. 1.1 Example of the use of the > operator and command concatenation

```
giuseppemandaglio@Mac testdir % ls
file1    file2    file3    file4
giuseppemandaglio@Mac testdir % ls>list.dat
giuseppemandaglio@Mac testdir % more list.dat
file1
file2
file3
file4
list.dat
giuseppemandaglio@Mac testdir %
```

cd -   Changes to the previous directory.

pwd   Shows the current directory.

1.3.1.7   Viewing Files

more filename   Displays files one page at a time. While the file is displayed, commands can be given:

space   Forward one page.

b   Back one page.

q   Quits viewing.

more *   Displays all files.

less file   Similar to `more`, but with more options.

cat filename   Displays the file.

tail filename   Shows the end of a file.

tail -100 filename   Last 100 lines.

head filename   Shows the beginning of the file.

wc filename   Counts lines, words, and characters in the file.

diff file1 file2   Differences between two files.

1.3.1.8   Creating, Copying, Deleting, Moving, and Zipping Files and Directories

touch filename   Creates an empty file or changes its date.

rm filename   Deletes a file. This command is very powerful and must be used with caution. Once a file or folder is deleted, it cannot be recovered. If you want to delete a folder and everything it contains recursively, you need to add the -r option to the rm command, where r stands for "recursive".

mv filename newname   Moves a file or renames it.

mv file1 file2 file3 directory   Moves a series of files to the directory.

mkdir namefolder   Creates a new folder.

cp file1 directory   Copy file1 in the directory.

tar   he tar command, which stands for tape archive, is used to create, modify, and extract archive files, often in combination with compression tools such as gzip.

gzip   gzip is used to compress files into an archive format, while its counterpart gunzip is used to decompress them.

cron   cron is used to automate the execution of commands or scripts at regular time intervals. It works in conjunction with the crond daemon. The crontab utility allows users to define and manage their own scheduled jobs.shell!cron

1.3.1.9 Links

ln -s real_file link_name   Creates a soft link.

ln real_file link_name   Creates a hard link.

1.3.1.10 File Permissions

chown user:group file   Changes the owner of a file.

chmod a+x file   Makes the file executable for everyone.

chmod og-r file   Only the owner can read the file.

chmod u-w file   The owner cannot modify it.

1.3.1.11 Finding Files: Find

Find Files by Name   : Example: find /path/to/directory -name "file.txt" Searches for files named file.txt in the specified directory and its subdirectories.

Find Files by Extension   : Example: find /path/to/directory -name "*.txt" Searches for all .txt files in the specified directory and its subdirectories.

Find Files by Type   : Example: find /path/to/directory -type d Searches for directories (-type d). Use -type f for files.

Find Files by Size   : Example: find /path/to/directory -size +50M Searches for files larger than 50 MB. Use -size -50M for files smaller than 50 MB.

Find Files by Modification Time   : Example: find /path/to/directory -mtime -7 Searches for files modified in the last 7 days. Use +7 for files modified more than 7 days ago.

Find Files by Permissions   : Example: find /path/to/directory -perm 755 Searches for files with 755 permissions.

Find and Delete Files   : Example: find /path/to/directory -name "*.tmp" -delete
  Searches for and deletes all .tmp files.
Find and Execute a Command   : Example: find /path/to/directory -name "*.log"
  -exec rm   Searches for all .log files and removes them.
Find Empty Files or Directories   : Example: find /path/to/directory -empty Searches
  for empty files and directories.
Find Files by User   : Example: find /path/to/directory -user username Searches for
  files owned by the specified username.

1.3.1.12   Finding Pattern in File: Grep

Basic Usage: The grep command searches for a specified pattern in a file or input.
  For example, grep "pattern" filename searches for the word "pattern"
  in the specified file and displays matching lines.
Case-Insensitive Search: The -i option allows for case-insensitive searches. For
  example, grep -i "pattern" filename finds "pattern", "Pattern", "PAT-
  TERN", etc.
Recursive Search: The -r or -R option allows grep to search directories recursively, finding all instances of a pattern in files within a directory and its subdi-
  rectories.
Show Line Numbers: The -n option displays the line numbers of matches within
  the files, making it easier to locate specific patterns in large files.
Invert Match: The -v option inverts the search, showing lines that do *not* match
  the pattern, which is useful for filtering out specific content.
Count Matches: The -c option displays the number of matching lines rather than
  the matching lines themselves, which can be useful for quick statistics.
Search for Whole Words: The -w option ensures that grep only matches whole
  words rather than substrings. For example, grep -w "word" filename
  will not match "wording."
Use with Regular Expressions: grep supports extended regular expressions, making it a flexible tool for pattern matching, data extraction, and text processing.

```
ps aux \textbar  grep firefox
grep -n -v pattern file   # Prints lines of file that
    do not (-v) contain the pattern, writes the line
    number (-n)
```

Code 1.1 Examples of command lines

File System Operations Disk names are:

/dev/sda   Primary master (/dev/sda1, sda2 etc. are the partitions).
/dev/sdb   Primary slave.
/dev/sdc   Secondary master.
/dev/sdd   Secondary slave.
/dev/nvme0n1   NVMe drive (common in modern systems).
/dev/fd0   The floppy.

These actions are permitted only to the root user.

fdisk -l /dev/sda   Shows the partition table.

fsck -v -r /dev/sda1   Repairs a file system.

mkfs -t ext4 /dev/sdb2   Creates an ext4 file system on the second partition of the primary slave.

mount -o remount,rw /dev/sda1   Remounts read-write.

Process and User Operations The interactive command interpreter offered in the Linux terminal offers many user commands that are useful to control computer activities, like checking the current working processes and the resources used, the used disk, managing the users' utilization rights, etc. A list of the most common user commands is as follows:

ps   Shows processes.

ps aux   Shows all processes.

du   Shows disk space used.

df   Shows file systems.

kill -9 process_number   Sends signal 9 to the process.

trap   Within a script, to react to a signal.

whoami   Shows the current user.

groups   Shows the groups the user belongs to.

passwd   Changes the password.

passwd username   Changes the user's password (only root or the user can do this).

members group   Shows members of the group.

who   Shows who is logged in.

w   Shows who is logged in and what they are doing.

last   Shows last logins.

useradd, userdel   Adds or deletes a user.

adduser, addgroup   In some distributions, these commands are more convenient than their counterparts.

date   Shows date and time.

date -s '+1 h'   Sets the clock forward by one hour (done by root).

uptime   Shows how long the system has been up.

uname -a   Shows operating system information.

uname -r   Shows kernel version.

lsof   Shows open files.

top   Shows system resource usage (exit with `Ctrl+C`).

free   Shows free memory.

times   Shows CPU usage.

time command   Executes and shows execution time.

nice command   Executes with low priority.

Batch Jobs

command   Executes the command.
command1 &   Runs the command in the background.
nohup command1 &   Continues running even when the terminal is closed.
at 8pm May 29 -f/home/user/program   Executes the program at 8 PM on May 29.
Ctrl+z   Interrupts the command.
jobs   Views the commands in progress.
bg %1   Sends job 1 to the background.
fg %1   Brings job 1 to the foreground.

Shell Variables

PATH   Where commands are searched for.
DISPLAY   Terminal for X11 output.
SHELL   Current shell.
USERNAME   User's name.
PWD   Current directory.
HOSTNAME   Hostname.
HOME   Home directory.
PS1   Prompt.

Conditional Command Execution

command && command2   command2 is executed if command succeeds.
command ||command2   command2 is executed if command fails.

1.4   Exercises

1.1 Practice using the cd command by navigating through your computer's folders and checking your current location.

1.2 Open a terminal and create a folder named test1. Then, move into that folder and create a new folder named test2. Navigate to test2 and ask the shell to display your current path.

1.3 Open a terminal and create a folder named test1. Then, run the ls command and copy all files with the .pdf extension into the test1 folder.

1.4 Use the grep command to check how often the cout statement appears in a C++ source file.

1.5 Write the ls command output to a file named test.txt, including the list of files and folders.

Chapter 2
Rudiments of Programming

2.1 Programming Base Languages Useful in Science

Programming is writing code(s)—a computer program—to tell a computer how to solve a particular problem or implement a particular task. A simple way of thinking about programming allows your computer to run the programs you use every day and your smartphone to run the apps you love: essentially, programming is a significant part of our world as we know it. In science, programming allows people and researchers to perform massive calculations, store data, and analyze results, automating processes that could be time-consuming, inaccurate, and tedious.

Many programming languages could be used in science. In this chapter, we chose two of the most important. We used open-source programming languages, **Python**—for its simplicity and the access to enormous opportunities offered by its libraries and packages—and **C++**—because it can be used at low levels (microcontroller programming) and high levels (object-oriented programming). In this book, we will mainly refer to code written for Linux, but all C++ and Python code can also be reproduced in other operating systems such as Windows and macOS. The Microsoft documentation Web page provides all the instructions for installing C++ and Python on this operating system. Similarly, Python and C++ can be easily installed by following the website guides.

Linux repositories are available in all its distributions. These repositories allow the installation of C++ (gnu-g++), Python3, and all the development libraries.

2.1.1 C++

C++ is the updated version of standard C, designed to include object-oriented programming. C++ code was born and mainly implemented to be compiled and run at the maximum speed possible. Still, existing C++ interpreters, like Cling of the analysis framework ROOT, allow running C++ code in small scripts like Python. C++ is known to have a steep learning curve due to its complexity and the need to gain a thorough understanding to use the language effectively. C++ requires a strong sense of fundamental and advanced programming concepts, such as memory management and object-oriented programming. The extensive standard library of C++, the extensive use of pointers, and the attention paid to performance and resource management are some of the most complex aspects of programming in C++. Those new to programming may find it difficult to learn C++, but mastering it provides a powerful tool for developing high-performance applications. C++ is used for operating systems, but Python is also implemented in C++. Despite the difficulties of learning C++, this skill is very useful in science, where one often needs to implement code with high speed (simulation and data analysis) or program firmware for micro-controllers, which is useful in constructing data acquisition systems. C++ code requires a compiler to be translated into machine language before it can be executed, or it can also be run using an interpreter. For example, the ROOT Data Analysis Framework provides an efficient C++ interpreter called CLing. Compiled C++ code is converted into machine language, allowing it to run at the maximum speed the hardware allows. In contrast, an interpreter translates and executes the code line by line, which results in slower performance. Interpreted code is typically suitable for small tasks or specific problem-solving, but is not ideal for large and complex projects.

There are several compilers for C++, each with its peculiarities, the best known and most used: **GCC (GNU Compiler Collection)**, it is one of the most popular compilers, available on many platforms, including Linux, macOS, and Windows (through MinGW). It supports different versions of the C++ standard. **Clang** is an open-source compiler part of the LLVM project. It is the default compiler in macOS, it is often used as an alternative to GCC on other platforms. **MSVC (Microsoft Visual C++)** is Microsoft provides the C++ compiler as part of Visual Studio. It is the standard compiler for developing Windows applications. **Intel C++ Compiler (ICC)** is a proprietary compiler provided by Intel. Optimized for Intel CPU architectures, often used in applications requiring high performance.

The source code can be written using any text editor. Linux offers many text editors, both with and without a graphical interface. The most common and popular ones are nano, pico, vim, emacs (without a graphical interface), or geany, Atom, kdevelop, and kate (with a graphical interface). On Windows, Dev-C++ represents a convenient choice, free and released under the GNU General Public License, while Xcode can be used under the OSX operating system for a similar reason.

2.1.2 *Python*

Python is a high-level programming language that is simple to implement and extremely powerful. It features an interpreter that makes it independent of the platform on which it is run. It can be used by typing instructions into the interpreter as if it were a calculator or by interpreting a series of instructions stored in a file (Python script). The extension for Python scripts is .py.

Python is introduced on its "About" webpage as follows: "*Python is powerful... and fast; plays well with others; runs everywhere; is friendly & easy to learn; is Open.*". In fact, Python is a high-level and modern programming language based on a simple and powerful syntax—similar to the English language—that allows developers to write programs with fewer lines than some other programming languages. Furthermore, Python works on different OSs (Windows, Mac, Linux, Raspberry Pi, etc.) and uses an *interpreter* that translates code of a high-level programming language into machine code line-by-line, granting the possibility to execute the code as soon as it is written. Python is an object-oriented programming language. Almost everything in Python is an object, with its properties and methods. The Python language allows you to develop more or less complex codes, even starting from the work of other programmers: this is the power of Open-Source and the Python developer community. Once a new library is created, the programmer can share it with everyone else. Libraries are very important because they collect functions and classes—we will describe them in detail later—constituting *modules*. Each module contains numerous functions that can be used for our purposes in the programming code.

Python's installation and first use are generally straightforward, and several Linux and UNIX distributions have already included a recent version of Python. If the user needs to download and perform a "fresh" installation of Python, its latest version can be downloaded from the "Download" webpage for several OSs—it is possible to install many other versions of Python, available on the webpage just reported.

Writing code in Python can be done using Integrated Development Environments (IDEs) and text editors such as Thonny, Pycharm, Netbeans, or Eclipse. These are particularly useful when managing larger collections of Python files.

A very good starting guide is given by the online documentation. It provides you with both basic and extensive information about Python and a description of Python's many libraries and syntax.

After installing the Python version and the IDE we chose, we can try to create a first and very simple example of a program that prints our name on the screen using the `print()` function. Once our IDE is started, we can type this simple line of code:

```python
print("My name is Sebastiano")
#Output: My name is Sebastiano
```

Code 2.1 First Python example with the print() function

and then run the code. In this example, note that we provide as input the print() function in Python that takes the text between quotes placed between the parentheses printing it to the screen as output.[1] Example 2.1 returns as output a *string*, one of the most important *data type* used in programming. Using the correct data type to

---

[1] Usually, in Python, if you want to print to the screen, the use of the print() function is recommended since it works both when writing in IDLE and with the text editor, but often in IDLE the argument is written directly without using print().

have a performing code and avoid making programming errors is essential. In this frame, we will introduce the most common base data types in the next section of this chapter.

2.2   Base Types

A very important part of understanding and/or writing a code program regards what types of data we use and how to handle this data. A **data type** is a classification of data that provides instructions to the compiler or interpreter on how to use the data. Data—as inputs and/or outputs—are processed using *bytes* that can be interpreted as representing values we understand, precisely as a basic data type. Most programming languages support data types, including integer, real, character or string, and Boolean. Common data types are reported in Table 2.1.

These common data types can usually be used in most programming languages, as well as additional complex and/or composite data types. As mentioned in Sect. 2.1, we will now see how to recall/employ these data types in Python and C++, also introducing some code examples.

2.2.1   C++

In C++, variables and their kinds must always be declared before use. If you declare the variables outside the `main` function, these variables are called *global variables*.

The global variables are visible to all functions present in the code. In contrast, if the variables are declared inside the main program or in other functions, these are local variables and live only in the function where they were declared.

Table 2.1 Common data types

Data type	Used for	Examples
Integer	All whole numbers[a] and their negative counterpart	7 , –12 , 0 , 999
Float	Fractional numbers[b]	3.14159 , –9.1 , 0.0
String	A sequence of alphanumeric characters[c]	Hello world! , Seb123
Boolean	Logical values[d]	True , False

[a] A collection of positive numbers without fractions and zero
[b] Numbers with a decimal point
[c] A combination of alphabetical and numerical characters
[d] Values indicating the relation of a proposition to truth

All variables must be declared. The following Code 2.2:

```cpp
#include <iostream>
int main()
{
  pi_greek=3.14;
  return 0;
}
```

Code 2.2 Example of C++ code with a mistake consisting in the use of an undeclared variable

will produce during compiling the following error:

```
gmandaglio75@buccellato:~> g++ prova.cpp
prova.cpp: In function 'int main()':
prova.cpp:6:3: error: 'pi_greek' was not declared in this scope
    pi_greek=3.14;
    ^~~~~~~~
```

That is why the variable *pi_greek* is not defined before; therefore, we cannot make the assignment instruction. The correct version of the Code 2.2 will be:

```cpp
#include <iostream>
int main()
{
  float pi_greek=3.14;
  return 0;
}
```

Code 2.3 Corrected version of Code 2.2

In this case, we defined the variable in the Code 2.3, and then in the same instruction, we did the assignment. It is possible, of course, to define the variable and then proceed with an assignment in another instruction. A variable is a word, a common word or one invented by the user, capable of storing information of the specified type according to its definition. In our example, the word pi_greek is a real variable of kind float storing the number 3.14.

In the Table 2.2 are reported the main kinds of data types in C++.

The main worry of a beginner in programming is remembering the names of variables to avoid mistakes in the compilation process, but it is more important to take care of the characteristics of the kind of variable to avoid logical mistakes. A common mistake, for example, is trying to store a real number in an integer variable; in this case, the decimal information will be lost.

2.2.2 Python

In Python, a data type can be identified as a class using the type() function. It can be applied to an argument returning the type of the given object:

Table 2.2 Main kinds of data types in C++

Groups	Names	Additional information
Character	char	1 byte, at least 8 bits
	char16_t	not smaller than Char, at least 16 bits
	char32_t	not smaller than char16_t, at least 32 bits
	wchar_t	greatest character supported
Signed integer	signed int	from -2147483648 to 2147483648 (32 bit)
	signed long int	from -2147483648 to 2147483648
	signed long long int	from -9223372036854775808 to 9223372036854775808
Real	float	from 1.17549×10^{-38} to $3.40282 \times 10^{+38}$
	double	from 2.22507×10^{-308} to $1.79769 \times 10^{+308}$
	long double	from 3.3621×10^{-4932} to $1.18973 \times 10^{+4932}$
Boolean	bool	assume only two value "true" or "false"
Void	void	No memory space used

```python
type(10)
#Returns <class 'int'> identifying the argument as an integer

type(10.56)
#Returns <class 'float'> identifing the argument as a float

type(True)
#Returns <class 'bool'> identifing the argument as a Boolean

type("My name is Sebastiano")
#Returns <class 'str'> identifing the argument as a string
```

Code 2.4 Type() function in Python

Note that combining the type() function with the print() one—see "Code 2.1"—provides the same results as in the example before:

```python
print(type(10))
#Returns <class 'int'> identifing the argument as an integer

print(type(10.56))
#Returns <class 'float'> identifing the argument as a float

print(type(True))
#Returns <class 'bool'> identifing the argument as a Boolean

print(type("My name is Sebastiano"))
#Returns <class 'str'> identifing the argument as a string
```

In Python, **integers** have no length limit—there is only a restriction due to the amount of memory your system has, so an integer can have as many digits as you need. Furthermore, it is possible to make quick calculations between integers again using the print() function. A few examples are reported in the following:

```
print(1083752935885029)
#Returns the value 1083752935885029

print(8375293 + 418397)
#Returns the value 8793690 performing the addition between
    two integers

print(8375293 * 418397 * 3)
#Returns the value 10512592395963, performing the
    multiplication between three integers

print(08375293 + 418397)
#Returns "SyntaxError", a syntax error in the programming
    code
```

Code 2.5 Integers in Python

Note that leading zeros in decimal integer literals are not permitted. If you want to check that the result of the operation is still an integer, use the type() function before the operation—the integer type is called int .

```
print(type(8375293 * 418397 * 3))
#Returns <class 'int'> confirming that the result of the
    operation is still an integer
```

A **float** is a floating-point number that is specified with a decimal point[2] A computer represents this type of number as binary fractions—e.g., the decimal fraction 0.237 has value 2/10 + 3/100 + 7/1000 (base 10), and in the same way, the binary fraction 0.001 has value 0/2 + 0/4 + 1/8 (base 2). Most base 10 fractions do not correspond exactly to base 2. So decimal floating-point numbers can be only *approximated* by the binary floating-point numbers stored in the machine. This base type is called float in Python. In science in general and physics in particular, this type of data is essential because there is often a need to indicate very small or vast numbers, which is done using *scientific notation*. Python allows writing huge and/or small numbers with the character e or E followed by a positive or negative integer. Examples of numbers are shown here in "Code 2.6".

---

[2] English-speaking countries (and China) use a decimal point and separate thousands with commas. Most European countries, such as Italy, use a space or a dot to separate thousands—but not necessarily when referring to millions—and a comma for decimals. Here we make use of the decimal point separating thousands with commas.

```
1   print(108.3752935885029)
2   #Returns the value 108.3752935885029
3
4   print(-5.)
5   #Returns the value -5.0
6
7   print(.6)
8   #Returns the value 0.6
9
10  print(3.1e8)
11  #Returns the value 310000000.0
12
13  print(-293.1 + 8397)
14  #Returns the value 8103.9 by doing the addition of two values
15  #In arithmetic calculations, note that the number written as
        an integer is considered a float too
16
17  print(-2.102e-3 * 2.5)
18  #Returns the value -0.005255, performing the multiplication
        between two floats
```

Code 2.6 Floats in Python

Tip

Generally, OSs use floating-point numbers in Python as 64-bit "double-precision" values, according to the IEEE 754 standard. In this frame, we have:

1. the maximum value for a float is about 1.8×10^{308} and Python will represent a number greater than this one by the string `inf`;

2. the number closest to zero is almost 5.0×10^{-324}, so any number closer to zero than that value is reported as `zero`.

A **string** is a sequence of *character* data—alphanumeric characters—and its type in Python is `str`. We already see this base type in the first Python code at the end of Sect. 2.1.2. Single or double quotes around strings are essential because they accurately mark the start and end of a string. Here are a few more examples of valid strings:

```
1 print('Hello world')
2 #Output: Hello world
3
4 print("Please, consider that the approximated value of pi is
      3.14159")
5 #Output: Please, consider that the approximated value of pi
      is 3.14159
6
7 print('')
8 #Returns an empty string
```
Code 2.7 Strings in Python

> **Tip**
>
> If you want to include a quote character as part of the string itself, the easiest
> ways are to delimit the string with the other type—if a string is to contain a
> single quote, delimit it with double quotes and vice versa—or by using triple-
> quoted strings.
>
> ```
> 1 print("This code contains 'Hello world' inside it!")
> 2 #Output: This code contains 'Hello world' inside it!
> 3
> 4 print('This code contains "Hello world" inside it!')
> 5 #Output: This code contains "Hello world" inside it!
> 6
> 7 print("'Test code with "Hello world" and 'Sebastiano'."')
> 8 #Triple-quoted string.
> 9 #Output: Test code with "Hello world" and 'Sebastiano'.
> ```

We can obtain even more complicated strings using (i) *escape sequences* in which
the use of a backslash ("\") character provokes its subsequent character sequence to
"escape" its usual meaning (e.g., the escape sequence "\n" doesn't return n as output,
but an ASCIILinefeed character), and/or (ii) *raw strings* that is preceded by "r" or
"R", which specifies that escape sequences in the associated string are not translated.

```
1 print("\u03C0 \u2192 \"pi\"\nValue: 3.14159...70679 \N{rightwards
      arrow} Approximated value: 3.14159\n\u03C0\\\u03C0 = 1")
2 #Escape sequence.
3 #Output: π → "pi"
4         Value: 3.14159...70679 → Approximated value: 3.14159
5         π\π = 1
6
7 print(r"\u03C0 \u2192 \"pi\"\nValue: 3.14159...70679 \N{
      rightwards arrow} Approximated value: 3.14159\n\u03C0\\\u03C0
      = 1")
8 #Raw string.
```

```
9 #Output: \u03C0 \u2192 \"pi\"\nValue: 3.14159...70679 \N{
     rightwards arrow} Approximated value: 3.14159\n\u03C0\\\u03C0
     = 1
```

Code 2.8 Complex strings in Python

The **Boolean** data type is either True or False and is called bool . When
we discuss expressions and their applications later, we will see the importance of
Boolean arithmetic and logic in investigating the truth or falsehood of an expression.

```
1 print(type(True))
2 #Output: <class 'bool'>
3
4 print(type(False))
5 #Output: <class 'bool'>
```

Code 2.9 Boolean data type in Python

Complex numbers are used a lot in science and are expressed in Python as a *real
part* plus an *imaginary part* represented as a number multiplied by j. This base type
is called complex :

```
1 print(-7+2j)
2 #Returns (-7+2j)
3
4 print(type(5.1-1j))
5 #Returns <class 'complex'>
```

Code 2.10 Complex numbers in Python

Note the importance of always inserting a value before "j"; otherwise, Python gives
an error.

2.3 Programming in C++ and Python

This section will present how to implement C++ and Python code, including the
code's structure, the use of libraries, variables and their operators, control structures
(sequence, selection, and repetition statements), functions, and classes.

2.3.1 *My First Script in C++*

The source code (which must be compiled before running) has to be written in a plain
text document, a file containing unformatted text. On Linux OS, many different text
editors can be used for this purpose: directly via terminal (vim, pico, nano, emacs,
etc.) or via a graphical interface (gedit, kate, atom, kdevelop, etc.). The same editors
can be installed on other operating systems, such as Windows and macOS. Of course,

these popular operating systems offer different robust native editor interfaces. To recognize what type of file it is, you need to look at its extension; in the case of a C++ source file, the extension is .cpp.

The following example reports the simplest code: the famous "Hello world," or, in other words, how to print a message in the terminal using a compiled program. The code allows us to give some instructions on the structure of a C++ program.

```
1 #include <iostream> //input/output library
2 int main(){  //main program
3 std::cout<<"hello world"<<std::endl;
4 return 0;
5 }
```

Code 2.11 Displaying the message 'Hello World' using C++

In the small piece of Code 2.11, the reason for the hardness in learning C++ is summarized. In only 4 rows of code, there is a lot of information that one has to know. First row, to have access to the operation of output cout it needs to include the library " iostream " by using the instruction #include <NameOfLibrary> ; second row, the main program is a *function*, the word int means that the function returns to the Operative System an integer value, the function has a name (in this case it is mandatory to be "main") and then two round brackets, if the brackets are empty means the function does not receive any information, all instructions of a function have to be enclosed in braces, each instruction finishes with a semicolon except for the inclusion on libraries. The first instruction of the *main* function contains the operator cout and its syntax to print a string on the screen, and the operator endl that allows to return a new line. The double slash is used to add a comment to the code. The comments are entirely transparent to the compiler. Adding comments to the code is warmly recommended, not only for sharing your code but also for your own good. Over time, it isn't very easy to remember the logic of one's code. The code finishes with the instruction return 0; , providing an integer value as stated before.

On Linux or macOS shell, it is enough to move in the folder containing the code and to write *g++ printhelloword.cpp*, the compiler checks the presence of errors and warnings and if mistakes are not present it compiles the code producing an executable file in binary (incomprehensible for human but clear for machine) and for default this file is "a.out". To run the file on the terminal it is enough to write ./a.out and the code will print on the shell the string "hello word".

Fig. 2.1 Python script execution

```
gmandaglio75@greyrock:~$ more Myfirstmacro.py
#this is the Python macro "Myfirstmacro.py"
print("hello")
gmandaglio75@greyrock:~$ python Myfirstmacro.py
hello
gmandaglio75@greyrock:~$ 
```

2.3.2   My First Script in Python

In Python, the code implementation is simpler. To repeat the exercise in Sect. 2.3.1, it is enough to write in an unformatted text file with the extension ".py" the function `print` as follows:

```
1  #this is the Python macro "Myfirstmacro.py"
2  print("hello")
```

Code 2.12   Python script "Myfirstmacro.py"

That's all! To add a comment in the Python macro, ignored by the interpreter, you must use the character # before the comment. The macro execution is shown in Fig. 2.1.

In this case, the only information you need to know is that to print a message on the screen, you must use the *print* function. This is the first example supporting that Python's learning curve is less steep than other programming languages, like C++.

2.3.3   Input/Output Operation

One of the primary purposes of programming code is to analyze data. To achieve this, it is essential to provide the tools necessary for communication with a human operator; all programming languages provide a comprehensive set of input/output (I/O) instructions, allowing programs to handle even large volumes of data efficiently. I/O operations encompass displaying messages on a screen, receiving input from the user, reading data from files or hardware devices, and writing output to files in a specified format. This section introduces the basic input/output functionality in C++ and Python.

2.3.3.1   C++

When we want to talk to someone, our first concern is understanding what they are saying and knowing how to speak to them. The same is true when we learn to code. These operations are called *Input/Output* operations. We already know the library furnishing the tool for these purposes, the `iostream` library, and how to include it in a code, and the output operator cout . The input operator is called cin . The syntax

of cin is very similar to that of cout: it is enough to remember that you must replace the minor-minor $<<$ operator of cout with a major-major $>>$. Writing code is the simplest way to familiarize yourself with these new operators. For example, a simple code can receive two numbers and return the sum. In the following, we report the Code 2.13:

```cpp
#include <iostream>
using namespace std;//allows us to avoid writing std: in
    front of the methods of iostream library: cout, cin, and
    endl
int main(){
    float firstNumber, secondNumber;//declaration of
    variables
    cout<<"please digit the first number you wish to sum \n";
     // \n inside the string works like endl operator
    cin>>firstNumber;//cin writes into the variable the value
     we wrote in the shell
    cout<<"Please enter the second number you wish to sum \n"
    ;
    cin>>secondNumber;
    cout<<"the sum of "<<firstNumber<<"+"<<secondNumber<<" =
    "<<fistNumber+secondNumer<<endl;
    return 0;
}
```

Code 2.13 Code able to sum two numeber

Let's briefly explain Code 2.13.

- The instruction in the second row allows us to avoid adding " std:: " in front of the methods cout, cin and endl.
- As previously reported, the variables must always be declared before use. In this case, the variables are of the "Real" data-type because we declared them float, as shown in row 4. We can use the comma to separate the variables in the case of multiple declarations.
- In row 5, note that the slash + n into the string in cout operator works like endl.
- Row 6 contains the first use in the code of the operator cin. cin write the value we wrote into the shell terminal in the variable, the same operation is repeated in row 8 *firstNumber*.
- cout can print multiple variables. As shown in row 9, we write (i) the string "the sum of", then (ii) the value of the variable *firstNumber*, (iii) the string "+", (iv) the value of the second number, (v) a string again, then (vi) the result of the sum of the two variables, and finally (vii) the command of the new row "endl".
- As already reported in Code 2.11, the function finishes in row 10 with the instruction return. Also, here, the function, according to its declaration, must return an integer value. For a convention, the value to be returned is zero if all parts of the code worked properly without any logic problem; in different cases, the pro-

grammer can decide to finish the program before the latest instruction returning a number different from zero (with the meaning of a problem).

As already seen, the input/output operations on the terminal are essential to make your code interactive, but to work on data, it is also necessary to manage the input/output operations on files. Generally, the data are often in text form (ASCIIfile), and reading and writing operations are fundamental. For these purposes, we introduce the use of the classes ifstream (reading class) and ofstream (writing class), which are available in the library fstream. It is easy to remember the functionality of the two classes; the initial letter of ifstream word i means input, while the o of ofstream means output.

The following Code 2.14 shows how ifstream works:

```
1  #include <iostream>
2  #include <fstream>
3  using namespace std;
4  int main()
5  {
6  ifstream readme;  //readme is an object
7  readme.open("/path/of/folder/nome_of_file.dat");//method open
       allows to access to the file in reading mode
8  float container;
9  leggimi>>container;  //the operator >> read a value of the
       file and assign it to the variable container
10 cout<<"that what we read "<<container<<endl;  //just to show
       it works
11 readme.close();//close method closes the link to the file
12 return 0;
13 }
```

Code 2.14 Example on how ifstream class works

and to follow an example on how the ofstream class works is reported in Code 2.15:

```
1  #include <iostream>
2  #include <fstream>
3  using namespace std;
4  int main()
5  {
6  ofstream writeme;  //readme is an object
7  readme.open("/path/of/folder/nome_of_file.dat");//method open
       allows to access to the file  in writing mode
8  float container =15.23;
9  writeme>>container;  //the operator >> write a value memorized
       into the variable container in the file
10 writeme.close();//close method closes the link to the file
11 return 0;
12 }
```

Code 2.15 Example on how ofstream class works

The objects of ifstream and ofstream, readme and writeme variables in Codes 2.14 and 2.15, work like cin and cout, but they act differently on files. The ifstream object reads the file using the operator $>>$, the reading flow starts from the first value from left to right, row by row, up to the end of the file. For each value successfully read, the ifstream object returns a logical true value to the program. When the last value is read, the further read operation gives the last value read. In such cases, the operation returns a logic false value; therefore, if one needs to reread the file, it is necessary to rewind it by using the methods clear() and seekg(), or trivially closing the file by the close() method and opening it again by open().

When we use the ifstream class, we must be careful and ensure that the object can connect to the file. This does not happen if the path and/or file name are incorrect. This is problematic because the compiler does not notice if the ifstream class object fails to open the file. The code can execute, but will provide obvious errors in the lucky cases or meaningless results in the worst cases. The class to avoid this problem provides the method is_open (), returning an actual logic value if the opening file procedure succeeded.

2.3.3.2 Python

When writing a Python program, the first thing to remember is that what is considered good practice in all programming languages, namely code indentation, is mandatory in Python. Indentation is what distinguishes one instruction from another.

If we wanted to write a program that prints the word "hello" on the screen, we would use the print() function, passing the string "hello" directly as an argument to the function, as shown in the Code 2.12.

Now we will use the learning of the print operation to address the indentation issue. The Python macro "Myfirstmacro.py" works properly, but if we wanted to repeat the same operation twice, we would repeat the same command. However, if we added a space in the second instruction, logically, we wouldn't be making any mistakes. However, Python doesn't work that way because doing so would violate its indentation rules, which are essential for determining where an instruction begins and ends, and the hierarchical structure that shows which instructions depend on others. In the following Code 2.16, we see an example of the Python syntax we are discussing:

```
1  #this is the Python macro "Myfirstmacro.py"
2  print("hello")
3   print("hello")
```

Code 2.16 Example of an indentation error in Python

In this case, Python would give an error message like this:

```
gmandaglio75@greyrock:~$ python buttami.py
  File "Myfirstmacro.py", line 2
    print("hello")
    ^
IndentationError: unexpected indent
```

To remove the mistake, it is enough to remove the additional space before the print() function in the second instruction of the code, and in this case, the code will print on the screen two times the word "hello" without any error message.

In addition to simple strings, it is also possible to print the contents of variables on the screen. To do this, pass these variables as arguments to the print function. If there are multiple contents to print, separate them with a comma. Below is a simple Python Code 2.17:

```python
#IPrintConstant.py
g = 9.81
pi = 3.14
print("A common physical quantity is worth ", g,"m/s^2 while
      a famous number is approximately",pi)
```

Code 2.17 Example of the use of print function in Python

The input() function handles input operations from the terminal. The function returns a string that can be stored in a variable. The function's argument can be a string in which the programmer can write a message to display on the screen. This feature can be used to ask the user for information to be used in the program.

Below, we implement a macro capable of asking a user for two numbers and returning their sum, printing the result on the screen. The macro will be deliberately "wrong" because it is important to realize that the input function does not read numbers or anything else; it reads strings. This is a repetition, but it is necessary to stress that, as in Code 2.18:

```python
# Ask the user for two numbers
num1 = input("Enter the first number: ")
num2 = input("Enter the second number: ")

# Attempt to sum the numbers (this will not work as expected)
sum_result = num1 + num2

# Print the result
print("The sum is:", sum_result)
```

Code 2.18 How really works the input function in Python

In this example reported in Code 2.18, the input function reads strings, not numbers, so the addition will concatenate the strings instead of performing a numerical sum. If we run the macro and provide, for example, the numbers 3.12 and 2.13 to the program, the program will print 3.122.13 on the screen. In this case, the program has correctly executed a string concatenation, a proper operation for many purposes, but not for what we wished. The way to correct the macro is to convert the strings into a number using the float function. The float function takes a string containing a number as an argument and returns a real number. Here is the correct version of the code:

```python
# Ask the user for two numbers
num1 = input("Enter the first number: ")
num2 = input("Enter the second number: ")

```

```python
5 # Attempt to sum the numbers (this will not work as expected)
6 sum_result = float(num1) + float(num2)
7
8 # Print the result
9 print("The sum is:", sum_result)
```

Code 2.19 How to read real numbers by using the input function, and printing the summing result in Python

Some thoughts on the Code 2.19 just written: we converted the string received into a real number, making the variable 'num1' a real number type variable. We then converted the string stored in the variable 'num2'. This shows us that Python, unlike C++, automatically understands the type of variable we are passing to it and therefore does not depend on the necessity of declaring variables before using them.

For those who love C++ and are tempted to think that this input function is a nuisance, let's try to immediately change their minds by implementing an exercise that is relatively simple with Python, but a little less so with other languages. Imagine asking a student to write down the grades he has obtained so far in his university career, separated by a space. Once acquired, we return the mean value, even if we do not know how many values the student entered.

```python
1 #Meangrades.py
2 grades = input("Please give me your grades")
3 grades =grades.split()
4 grades = list(map(float, grades))
5 print ("the mean value of your grades is  ", sum(grades)/len(
      grades))
```

Code 2.20 Example of the use of split method and the map, list, sum, len functions in Python

In the Code 2.20, we introduced the split method of the string class, the accessory functions map, list, sum, and len. The code is short, but we can make it shorter in the following way:

```python
1 #Meangrades.py
2 grades = list(map(float, input("Please give me your grades").
      split()))
3 print ("the mean value of your grades is  ", sum(grades)/len(
      grades))
```

Code 2.21 Compact version of Code 2.20

The code in Listing 2.21 is brief but rich in meaning, as becomes clear when we analyze it step by step. The variable grades stores a string containing all the score values entered by the user via the terminal, captured using the input function. The split method of the string class then transforms this string into a list of substrings. Finally, the map() function converts these substrings into floating-point numbers. Notice how powerful the code is: its functionality doesn't depend on the student's identity or the stage of their academic career, as it effectively solves the problem of determining the number of exams passed. We encourage the reader to repeat the exercise in C++, either by handling the input as a single string, as we did in Python, or by passing the score values directly as integers.

The open class is used to read a file. It needs a string containing the path of the file plus its name. In Code 2.22, the input function is used to ask the user to enter the file's path and to read it, storing it in the variable *name*. The *open* object returns a string containing everything written into the file using the read() method. If the student stored the achieved grades in a file entitled mymark.txt, we can rewrite the Code 2.21 as follows in Code 2.22, obtaining the same functionality.

```
#Meangrades.py
name = input("Please, tell me how to reach the file, path
    plus name, or name: ")
grades = list(map(float, open(name).read().split()))
print ("the mean value of your grades is  ", sum(grades)/len(
    grades))
```

Code 2.22 Version of the Code 2.20 reading data from file

2.3.4 *Multidimensional Variables and Operators*

This paragraph introduces multidimensional variables, their definition and use, and the operators that allow actions on variables. The purpose of any computer program is to process information, and variables and operators are the fundamental elements that make this possible. The topics will be presented by drawing a parallel between the C++ and Python languages.

2.3.4.1   Array

The simplest type of variable that allows managing data of the same kind, such as numbers or characters, by simply using the variable type, the variable name, and an index to navigate through the different stored elements, is called an array. In the paragraph, we start with the simplest example of a data collection, the array, up to more complex structures like list, tuple, dictionary, etc.

C++

To define an array, it is necessary to declare what kind of data we have to store in the array variable, the name of the variable, and the dimension of the array by using an integer enclosed in two square brackets. If we want to store ten real numbers of floating point kind in an array, we will write:

```
#include <iostream>
int main()
{
  float myarray[10]; //10 in this case define the dimension
  myarray[0] = 4.2; //assigning a value to the first element
     of the array
```

```
6   myarray[1]=12.4;//assigning a value to the second element
       of the array
7   myarray[9] = 20.1;//assigning a value to the last element
       of this array
8   return 0;
9 }
```

Code 2.23 C++ array definition example

In Code 2.23, we note that similarly to simple variables, we must declare the kind of array variables, in this example a float, then the name of the array variable, and finally the dimension. The dimension of the array is settled during the definition and it is not modifiable later, the index used to access the element of the array varies between zero, the first element, and the dimension of the array minus one (because the numbering of the component in c++ starting from zero and not one). The array discussed in the examples reported in Code 2.23 has only one dimension, but it is possible to define an array with multiple dimensions. For example, to determine an array with two dimensions, a matrix, it is enough to add another double square bracket in the definition of the variable, including an integer number defining the dimensions, as in Code 2.24.

```
1 #include <iostream>
2 int main()
3 {
4   float mymatrix[10][5]; //in declaration 10 and 5 represent
       the dimension
5   mymatrix[3][2] = 14.23; //14.23 is assigned to the element
       of the two-dimensional array (3,2)
6   return 0;
7 }
```

Code 2.24 C++ two-dimensional array definition example

Python

In Python, arrays store multiple values in a single variable. However, Python does not have built-in support for arrays like some other programming languages. Instead, you can import libraries like `array` or NumPy[3] for array functionalities.

Here, we report the `array` module included in Python's standard library, so no additional installation is required. It provides a way to create arrays with elements of the same type:

```
1 import array as arr
2 a = arr.array('i', [1, 2, 3])
3 print(a)
4 # Output: array('i', [1, 2, 3])
```

Code 2.25 Python array library example

---

[3] Note that we will talk extensively about NumPy and its modules in Chap. 5.

2.3.4.2   List

C++

Lists are sequence containers that support non-contiguous memory allocation. Unlike vectors, lists have slower traversal, but insertion and deletion are efficient once a position is located (constant time). Typically, we refer to a List as a doubly linked list. To implement a singly linked list, we use a `forward_list`.

The `std::list` class represents the List container. It is part of the C++ Standard Template Library (STL) and is defined in the '<list>' header file.

```cpp
#include <iostream>
#include <list>

int main() {
    std::list<int> mylist = {1, 2, 3, 4};
    for (int item : mylist) {
        std::cout << item << " ";
    }
    return 0;
}
```

Code 2.26 C++ list example

Basic operations on lists with C++ comprehend:

- Adding elements:

  - `push_front()` : Adds an element to the beginning of the list.

  - `push_back()` : Adds an element to the end of the list.

- Remove elements:

  - `pop_front()` : Removes the first element of the list.

  - `pop_back()` : Removes the last element of the list.

- Accessing elements:

  - Iterators and Indexing (see the following example).
  - `front()` : Returns the first element of the list.

  - `back()` : Returns the last element of the list.

```cpp
#include <iostream>
#include <list>

int main() {
    // Define a list and print it
```

```cpp
std::list<int> mylist = {1, 2, 3, 4};
std::cout << "The elements of the list are: ";
for (int item : mylist) {
    std::cout << item << " " ;
}

std::cout << '\n';

//Print the size of the list
std::cout << "The size of 'mylist' is: " << mylist.size()
 << " " << std::endl; // Output: 4

// Add 6 to the end of the list
mylist.push_back(6);
std::cout << "Adding '6' to the list..." << std::endl; //

//Print the elements of the list
std::cout << "The elements of the list are: ";
for (int item : mylist) {
    std::cout << item << " " ;
}

std::cout << '\n';

//Print the new size of the list
std::cout << "The new size of 'mylist' is: " << mylist.
size() << " " << std::endl; // Output: 5

std::cout << '\n';

//Print the last element of the list
std::cout << "The last element of the list is: " <<
mylist.back() << " " << std::endl;    // Output: 6;

// Create an iterator to point to the first element of
the list
std::list<int>::iterator itr = mylist.begin();

// Increment itr to point to the 2nd element
++itr;

//Display the 2nd element
std::cout << "Second element of 'mylist': " << *itr <<
std::endl;

return 0;
}
```

Code 2.27 Basic operations with C++ list

This example includes <iostream> and <list> libraries for input/output operations
and using lists, respectively. In the main function, we create a list of integers 'mylist'
with elements 1, 2, 3, and 4, and print them with the actual list size —which is 4. By
using the command   mylist.push_back(6) , it is possible to add the element 6 to the

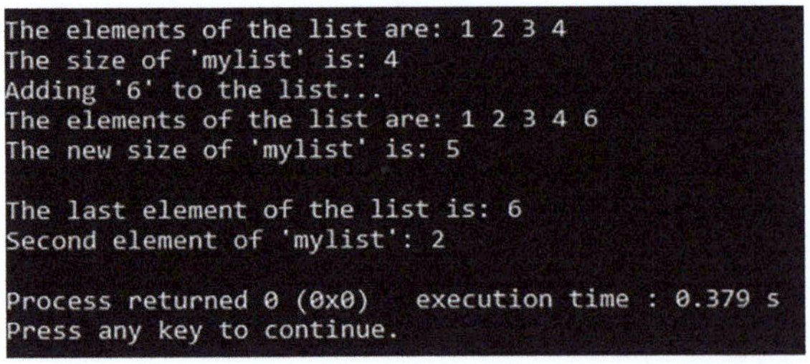

Fig. 2.2 Output of Code
2.27

```
The elements of the list are: 1 2 3 4
The size of 'mylist' is: 4
Adding '6' to the list...
The elements of the list are: 1 2 3 4 6
The new size of 'mylist' is: 5

The last element of the list is: 6
Second element of 'mylist': 2

Process returned 0 (0x0)    execution time : 0.379 s
Press any key to continue.
```

end of the list. This operation is confirmed by new prints of (i) the updated elements of the list, which are now 1, 2, 3, 4, and 6, and (ii) the new size of the list, which is 5. The part of the code in which we make use of   `mylist.back()` and   `std:cout`

allows us to print the last element of the list. Finally, the last part of example 2.27,

```cpp
std::list<int>::iterator itr = mylist.begin();
++itr;
std::cout << "Second element of 'mylist': " << *itr << std::endl;
```

creates an iterator 'itr' that points to the first element of the list, increments it to point to the second element, and prints the second element, which is 2. The result of this code is reported in Fig. 2.2.

Python

In Python, a list is a versatile data structure that allows you to store multiple items in a single variable. Lists are a fundamental part of Python and are used extensively due to their flexibility and ease of use. The items in a list have a defined order, which will not change unless explicitly modified. Furthermore, they are mutable, meaning you can change, add, or remove items after creating the list, even allowing duplicate values. List items are indexed, with the first item having an index of 0, the second item an index of 1, and so on.

As illustrated in the following example, a list can be created by placing all the items (elements) inside square brackets "[ ]", separated by commas.

```python
mylist = ["blue", "red", "green"]
print(mylist)
# Output: ['blue', 'red', 'green']
```

Code 2.28 Create a Python list

Accessing list items can be done by referring to the index number.

```python
print(mylist[0])
# Output: blue
```

Code 2.29 Access list items with Python

Since lists are mutable, we can modify the value of a specific item by referring to its index.

```python
mylist[2] = "yellow"
print(mylist)
# Output: ['blue', 'red', 'yellow']
```

Code 2.30 Change list items with Python

Another essential feature is the addition of items in a list using methods like
append() , insert() , or extend() .

```python
mylist.append("orange")
print(mylist)
# Output: ['blue', 'red', 'yellow', 'orange']
```

Code 2.31 Add list items in Python

We can remove items from a list using methods like remove() , pop() , or
clear() .

```python
mylist.remove("yellow")
print(mylist)
# Output: ['blue', 'red', 'orange']
```

Code 2.32 Remove list items in Python

Python lists come with a variety of built-in methods, such as sort() , reverse() ,
copy() , and more.

```python
mylist.sort()
print(mylist)
# Output: ['blue', 'orange', 'red']
```

Code 2.33 Example of the sort() method in Python

2.3.4.3 Dictionary

Python

In Python, a dictionary is a collection of key-value pairs. It is an unordered, mutable,
and indexed data structure that efficiently stores and retrieves data. A dictionary can
be created by using curly braces "{ }" or the dict() function:

```python
# Using curly braces
# Using curly braces
my_dict_curly = {
    "physical_quantity": "mass",
    "value": 30,
    "unit": "kg"
}

```

```
 9 # Using the dict() function
10 my_dict_func = dict(physical_quantity = "mass", value = 30,
       unit = "kg")
11
12 print(my_dict_curly)
13 print(my_dict_func)
14
15 # Output: {'physical_quantity': 'mass', 'value': 30, 'unit':
       'kg'}
16 # Note that the two print commands provide the same output.
```

This simple example shows how to create an equal dictionary using the two methods mentioned above. Inside it are 3 key-value pairs, e.g., "physical_quantity" as key and "mass" as its associated value, and so on.

It is possible to access values in a dictionary by using their corresponding keys

```
1 ...
2 print('The ' + my_dict_func["physical_quantity"] + ' has been
       measured giving ' + str(my_dict_func["value"]) +
       my_dict_func["unit"] + ' as result.')
3
4 # Output: The mass has been measured, giving 30kg as a result
       .
```

Similarly, the user can add or update entries (key-value pairs) with the following commands:

```
1 ...
2 # Adding a new key-value pair
3 my_dict_func["measurement(s)"] = 10
4
5 # Updating an existing key-value pair
6 my_dict_func["unit"] = "g"
7
8 print(my_dict_func)
9 # Output: {'physical_quantity': 'mass', 'value': 30, 'unit':
       'g', 'measurement(s)': 10}
```

Instead, if you do not need key-value pairs, you can make use of the "del" statement or the `pop()` method to remove entries.

We report some useful dictionary methods in Table 2.3.

The following example shows the use of a dictionary:

```
1 # Creating a dictionary
2 my_dict_func = dict(physical_quantity = "mass", value = 30,
       unit = "kg")
3 print(my_dict_func)
4 # Output: {'physical_quantity': 'mass', 'value': 30, 'unit':
       'kg'}
5
6 # Accessing values
7 print(my_dict_func["unit"])
8 # Output: 30
9
```

Table 2.3 Dictionary methods in Python

Module	Functioning
.keys()	Returns a view object of the dictionary's keys
.values()	Returns a view object of the dictionary's values
.items()	Returns a view object of the dictionary's key-value pairs
.get(key)	Returns the value for the specified key if the key is in the dictionary
.clear()	Removes all items from the dictionary

```python
10  # Adding a new key-value pair
11  my_dict_func["measurement(s)"] = 10
12
13  # Updating an existing key-value pair
14  my_dict_func["unit"] = "g"
15
16  print(my_dict_func)
17  # Output: {{'physical_quantity': 'mass', 'value': 30, 'unit':
        'g', 'measurement(s)': 10}
18
19  # Removing a key-value pair
20  del my_dict_func["measurement(s)"]
21
22  # Iterating through key-value pairs
23  for key, value in my_dict_func.items():
24      print(f"{key}: {value}")
25  # Output: physical_quantity: mass
26  #         value: 30
27  #         unit: g
```

Code 2.34 Example of creating and using dictionaries in Python

C++

Creating a dictionary in C++ can be done using the std::map or std::unordered_map from the Standard Template Library (STL). As shown in paragraph 2.3.4.3, these containers store key-value pairs and allow for efficient retrieval of values based on their keys. A simple example—reporting some units of measurement in the International System of Units (SI)—is given here:

```cpp
1  #include <iostream>
2  #include <map>
3  #include <string>
4
5  int main() {
6      // Create a map to store dictionary
7      std::map<std::string, std::string> dictionary;
```

```
 8
 9     // Insert key-value pairs into the map
10     dictionary["newton"] = "The unit of force in the SI.";
11     dictionary["ampere"] = "The unit of electric current in
       the SI.";
12     dictionary["second"] = "The unit of time in the SI.";
13
14     // Access and print values using keys
15     std::cout << "Newton: " << dictionary["newton"] << std::
       endl;
16     std::cout << "Ampere: " << dictionary["ampere"] << std::
       endl;
17     std::cout << "Second: " << dictionary["second"] << std::
       endl;
18
19     return 0;
20 }
```

Code 2.35 Example of creating and using dictionaries in C++

In this example, making use of necessary headers (e.g. iostream and map), we create
a `std::map` named "dictionary" to store our key-value pairs. After that, we insert
some key-value pairs into the map and then access and print the values using their
corresponding keys.

2.3.4.4   Arithmetic Operators in C++ and Python

The arithmetic operators are used to implement the algebra between the code variables; the result of each arithmetic operation is a number, and they do not work with
character variables. The operators are reported in Table 2.4. The hierarchy in the
order of operations is that of standard mathematics. Note that the division of two
integer numbers is an integer number; therefore, if the integer divider is higher than
the dividend, the result is zero.

Other helpful arithmetic operators are the increment and decrement operators,
available only in C++. To increment a variable in Unity, it is enough to implement
an instruction with the variable's name followed by a double plus symbol, and for
the decrement, followed by a double minus symbol.

Table 2.4 Arithmetic operators in C++

Operator	Operation
+	Sum
−	Subtraction
*	Multiplication
/	Division
%	Modulo operation

Table 2.5 Assignment operators

Operator	Use	Meaning
=	a = b;	a assume the value of b
+=	a += b;	a = a +b;
-=	a -= b;	a = a -b;
*=	a *= b;	a = a *b;
/=	a /= b;	a = a / b;
%=	a %= b;	a = a % b;

Increment and decrement operators in C++

```
int a = 10;
int b = a;
a++; //it is equivalent to a = a +1;
b--; //it is equivalent to b = b -1;
```

In the present examples, a++ means that the variable a changes value from 10 to 11, while b–means that b equals nine after the instruction.

2.3.4.5 Assignment Operators in C++ and Python

The assignment operators are necessary to assign a value to a variable, the list is reported in Table 2.5.

2.3.4.6 Relational and Logic Operators in C++ and Python

The relational operators allow us to establish the relationship between two variables; the operation returns a logical value "true" if the relation is satisfied, "false" in the opposite case. Table 2.6 reports a list of the relational operators.

Relational operators can be used to construct logic statements whose truth value can be used to control a program's behavior. It is possible to link more logic statements using the logical operator, as reported in the Table 2.7 for C++.

In Python, unlike C++, the operator AND is *and*, the OR is *or*, and the NOT is *not*. Therefore, if we wish to link the statements $a < b$ and $b > c$ with the OR operator in Python, we must write $a < b$ *or* $b > c$. A common programming mistake is using

Table 2.6 Relational operators

Name	Operator	Use	Meaning
equality	==	(x==y);	returns true if x and y are equal, false if not
disequality	!=	(x!=y);	returns true if x and y are different, false if not
majority	> (>=)	x>b;	returns true if x is higher (higher-equal) to y, false if not
minority	< (<=)	x<b;	returns true if x is lower (lower-equal) to y, false if not

Table 2.7 Logical operators

Name	Operator	Use	Meaning
AND	&&	(x>a) && (x<b);	returns true if both expressions are true
OR	||	(x<a) || (x>b);	returns true if at least one expression is true
NOT	!	!(a<b);	Invert the logical result of the expression

the assignment operator = with the equality ==. This error is annoying because the compiler does not detect it, as it's not a syntax error. It becomes evident only when the code does not work as expected.

2.3.5   *Control Structures*

Control structures are the fundamental tool for building a logical algorithm. Without them, elementary operations become complex or unfeasible. Imagine, for example, implementing a summation of numbers or a calculator providing the five arithmetical operations. Control structures can be sequence statements (if, if—else, switch) or repetitive statements (while, do-while, for). All programming languages, with slightly different syntax, use control structures.

2.3.5.1   If, If-Else, and Switch Operators in C++

The syntax of the if-operator requires the keyword *if*, followed by a logical expression enclosed in round brackets, and then the instruction or block of instructions to be executed. A single instruction can follow the condition directly, but it is recommended always to use curly brackets to enclose the instructions, even if there is only one. The instructions within the if block are executed only if the condition inside the brackets evaluates to true; otherwise, they are skipped (see Listing 2.36).

```
if(conditions){
    statement 1;
    statement 2;
    ....
}
```

Code 2.36 If-operator operating structure

The if-operator, in combination with the else-operator, works as a switch with two possibilities. If the condition is *true*, it allows the execution of the instructions following the if-operator; otherwise, it executes the instructions that follow the else-operator, skipping the first.

These operators can be nested, allowing for more complex logical structures within the code. For example, consider the case of a second-order equation, where the nature of the solutions depends on the value of the discriminant (see Listing 2.37).

```
//known the coefficients of equation a*x^2 + b*x + c = 0
float delta = b*b - 4*a*c; //define the determinant of the
    equation

if(delta > 0 ){
cout <<"the first root is equal to "<<(-b + sqrt(delta))/(2*a
    )<<endl;
cout <<"the second root is equal to "<<(-b - sqrt(delta))/(2*
    a)<<endl;
}
else{
    if(delta==0){
        cout<<"the only solution is "<< -b/(2*a)<<endl;
    }
    else{
        cout<<"the equation does not have any real solutions"
    <<endl;
    }
}
```

Code 2.37 Decisional code in a second order equation

The following example, shown in Code 2.38, demonstrates how the if operator can be used to check whether an ifstream object is successfully linked to a file, preventing potential issues.

```
#include <iostream>
#include <fstream>
using namespace std;
int main()
{
ifstream writeme;
readme.open("/mypath/nome_of_file.dat");
```

```
 8  if(!readme.is_open()){
 9  cout<<"sorry! I'm unable to reach the file"<<endl;
10  }
11
12  float container =15.23;
13  writeme>>container; //the operator >> write a value memorized
            into the variable container in the file
14  writeme.close();//close method closes the link to the file
15  return 0;
16  }
17
18  }
```

Code 2.38 Decisional code in a second order equation

The switch operator is structured by a control value and a list of key values; the operator allows the execution only for the instructions following the key value equal to the control one (see the example in Code 2.39)

```
 1  /*
 2  switch(variable of Control value) {
 3     case key value:
 4        // instructions
 5     case key value:
 6        // instructions
 7  */
 8  float a =5, b=7;
 9  cout<<"let me know if you want to perform a sum, difference,
           multiplication or division"<<endl;
10  string choice;
11  cin>>choice;
12  switch(choice){
13     case "sum":
14     cout<<"the result of the sum is "<< a+b;
15     case "difference":
16     cout<<"the result of the sum is "<< a-b;
17     case "multiplication":
18     cout<<"the result of the sum is "<< a*b;
19     case "division":
20     cout<<"the result of the sum is "<< a/b;
21  }
```

Code 2.39 Switch operator

2.3.5.2   If, and Logic Operators in Python

The conditional operators used in Python have the exact keywords as in C++: if and else. In contrast, the switch operator does not exist, and one must implement its functionality using different strategies. The if-else statement in Python is made by the keyword **elif**. In the following Code 2.40, we implement a Python script that asks a user for the coefficients of a second-order equation and gives the solutions.

```python
import math
coeff = input("please, tell me the coefficients of 2nd order
    equation ax^2+bx+c \n")
coeff = list(map(float,coeff.split()))
delta = coeff[1]*coeff[1] - 4*coeff[0]*coeff[2];
if delta > 0 :
    print("the first root is equal to ",(-coeff[1] + math.
    sqrt(delta))/(2*coeff[0]))
    print("the second root is equal to ",(-coeff[1] - math.
    sqrt(delta))/(2*coeff[0]))
elif delta == 0:
    print("the only solution is ", -b/(2*a))
else :
    print("the equation does not have any real solutions")
```

Code 2.40 Solution of a second-order equation in Python

By looking at the Code 2.40, we note that on line 1 we imported the math library
because we need to use the square-root function, on line 2 we end the string passed
to the input function with \n to force the new line on the terminal during running
the script, on line 5 we see the construction of the if-statement. The construction of
an if statement begins with the keyword if, followed by a logical condition (without
parentheses), and a colon that introduces the statement's body. The logical operators
used are the same as in C++. The else statement follows similar rules but does
not require a condition, just as in C++. The elif statement is particularly useful
for handling multiple conditions. Without it, one would need to nest multiple if
statements, which can make the code less readable (see Code 2.41).

```python
import math
coeff = input("please, tell me the coefficients of 2nd order
    equation ax^2+bx+c \n")
coeff = list(map(float,coeff.split()))
delta = coeff[1]*coeff[1] - 4*coeff[0]*coeff[2];
if delta > 0 :
    print("the first root is equal to ",(-coeff[1] + math.
    sqrt(delta))/(2*coeff[0]))
    print("the second root is equal to ",(-coeff[1] - math.
    sqrt(delta))/(2*coeff[0]))
    if delta == 0:
        print("the only solution is ", -b/(2*a))
    else :
        print("the equation does not have any real solutions"
    )
```

Code 2.41 Solution of a second-order equation in Python, by using only if-statement

The if-statement on line 8 of Code 2.41 is nested to the one on line 5.

The switch-statement can be implemented as a sequence of if - elif - elif - ...
- else, or by using the **match** and **case** keywords starting from Python version 3.10.
In the Code 2.42, we rewrite the Code 2.39 with the match-statement.

```python
numbers = input("I'm a calculator program...give me two
    numbers \n")
numbers= list(map(float,numbers.split()))
```

```python
3 choice = input("let me know if you want to perform a sum,
      difference, multiplication or division")
4 match  choice:
5     case "sum":
6           print("the result of the sum is ", a+b)
7     case "difference":
8           print("the result of the sum is ", a-b)
9     case "multiplication":
10          print("the result of the sum is ", a*b)
11    case "division":
12          print("the result of the sum is ", a/b)
```

Code 2.42 Solution of a second-order equation in Python, by using only if-statement

In the example implemented, we use a single condition after if- and elif-operators. However, how can we proceed if we combine more sentences to be evaluated by the conditional operators? We need to use the logic operators, and in Python are implemented as already said by the keywords **and**, **or**, and **not**. The first two combine two conditions, and the latter changes the opposite of the logical value of a sentence or a boolean variable. The Code 2.43 is reported as an example using the and operator.

```python
1 word = input("digit a password longher than 10 characters,
      not exceeding 20")
2 if len(word) >=10 and len(word) <=20:
3     print("the password lenght is correct")
4 else:
5     print("the password lenght is uncorrect, try again!")
```

Code 2.43 Example of and operator use in Python

2.3.5.3   Loop: While, Do-While, for Operators in C++

Whenever a set of instructions needs to be executed repeatedly, a loop should be used. The Loop can be implemented by using the *while*, *do-while* or *for* operators. The *while* and *do-while* operators are used when you do not know a priori how many repetitions you have to perform; the for operator is used when you know how many there are.

C++

The operation of the while loop is summarized below:

```cpp
1 while(condition){ //condition can be the result of a
      comparison, boolean variables, or logic combination of
      them
2 instruction 1;
3 instruction 2;
4 instruction 3;
5 .....
6 }
```

Code 2.44 While loop scheme

The while loop, similar in structure to the if statement, executes its block of instructions—enclosed in curly brackets—repeatedly as long as its condition remains true. This type of loop can run indefinitely (an infinite loop) if the condition continually evaluates to true, or it can terminate when an instruction inside the loop alters the condition to false. An endless loop may indicate a programming error without a proper exit strategy. Still, in some cases, it is intentional, for example, when a system must continuously monitor peripherals, such as in a home security system or an environmental sensing device.

In the Listing 2.45 is reported, as an example, the summation of the numbers between 0 and 9. Please note that if we swap the instruction of line 5 with the one of line 6, we also include in the summation the number 10, which is not what we wanted. We'll repeat this statement many times in this book: the order of the instructions is crucial for the correct logical functioning of the program. Practicing how to write a program's flowchart is very helpful for understanding the importance of the order of instructions.

```cpp
//repeats instructions ten times, sum the numbers 0-9
int count = 0;
int sum =0;
while(count<10){
sum+=count;
cout++;
}
```

Code 2.45 Summation example with while-loop

Looking at these few examples, we can note that the while loop always needs a true value to execute the loop's body for the first time. This could be desired and necessary in many circumstances. In cases where we want to execute the loop instructions for the first time without evaluating the trueness of any condition, we can use the do—while loop. In the listing 2.46, we rewrite the code of listing 2.45 by using the do-while structure. Note that at the end of the loop, after the while, we close the do-while loop with a semicolon.

```cpp
//repeats instructions ten times, sum the numbers 0-9
int count = 1;
int sum =0;
do{
sum+=count;
cout++;
}while(count<10);
```

Code 2.46 Summation example with do-while-loop

As an example of the application of a do-while loop, we report the following code:

```cpp
//The program asks a user two numbers and tells how is larger
#include <iostream>
#include <string>
using namespace std;
int main(){
```

```
 6   string decision;
 7    do{
 8        float first, second;
 9        cout<<"please give me two real numbers"<<endl;
10        cin>>first>>second;
11        if(first>second){
12           cout<<"the first number is larger than the second"<<
      endl;
13        }
14        else{
15           cout<<"the second number is larger than the first"<<
      endl;
16        }
17    cout<<" Do you want to compare another two numbers?"<<endl;
18     cin >>decision;
19     while(decision != "yes" && decision != "not"){
20         cout<<"Sorry, I do not understand! Please answer yes or
        not"<<endl;
21         cin>>decision;
22      }
23    }while(decision =="yes");
24
25    return 0;
26  }
```

Code 2.47 Program telling an user what number is the larger one

The program reported in listing 2.47 asks two numbers and estimates which one is larger. Note the use of a do-while loop, which ensures the loop's instructions are executed at least once; the if-else structure, which determines the larger number; and a while loop, employed to prevent ambiguous responses.

As already said, the *for* loop is used when we know the number of repetitions a priori. The operating mode is reported in the following scheme:

```
 1  //The program tells a user which number is larger
 2  #include <iostream>
 3  using namespace std;
 4  int main(){
 5   //sum the integer form 0 to 100
 6  int sum =0; //initialize the sum to zero
 7  // for( start condition; stop condition; increment)
 8   for(int i=0; i<=100; i++){
 9  sum+=i;
10   }
11   cout<<"the sum to the first 100 numbers is "<<sum <<endl;
12   return 0;
13  }
```

Code 2.48 Summation example with for-loop

The *for* loop needs an initial condition for a variable, a stop condition, and an increment for the variable. Note that a semicolon separates each instruction. The variable in a for loop is generally an integer and acts as a counter, but it can also be a real (floating-point) variable. The for loop variable can be used inside the loop, for example when generating numerical sequences, or it may not take part in the repeated operations within the loop at all.

Loop: while, do-while, for operators in Python

The loop operators in Python have the same names as in C++, with little operational differences. The *for* loop operator, for example, works as shown in Code 2.49:

```python
#Python script - for-operator example
things = [0,1,2,3,"hello",5,6,7,8,9]
for var in things:
    print(var)
```

Code 2.49 Example of Python code using for-loop

The loop is written with the keyword "for" preceding a variable, the keyword "in", a list, and a semicolon. After the semicolon, the instructions depending on the loop must be indented to tell Python that the instruction belongs to it. In the example, the script prints the list "things" on the screen using the variable "var", sequentially assuming the value of each list element.

We can use the Python built-in function range() to generate an integer number sequence in a loop. The purpose of the range() function in the loop is to produce a numerical list. In particular, when provided with a value N, the range function produces a sequence of numbers starting at zero and ending at N-1, incrementing by 1. The following Code 2.50 prints the numerical sequence from zero to 9, printing 4 instead of the "hello" of the previous example.

```python
#Python script - for-operator example
for var in range(10):
    print(var)
```

Code 2.50 Example of Python code using for-loop and the function range

The *for* operator in Python is more flexible; it can loop on all kinds of lists and a string, as in Code 2.51. In the following example, the code prints all characters into the string, a list of characters.

```python
#Python script - for-operator example
words = "hello"
for var in words:
    print(var)
```

Code 2.51 Example of Python for-loop looping on a sting variable

and the result is:

```
h
e
l
l
o
```

If we transform with the method split() a string into a list of strings before looping, the Code 2.52 will be:

```python
#Python script - for-operator example
words = "hello, I'm Python and I'm very flexible!"
for var in words.split():
    print(var)
```

Code 2.52 Example of Python for-loop looping on a sting containing a sentence splitter in single word

and the results will be

```
Hello,
I'm
Python
and
I'm
flexible!
```

To introduce the while loop operator in Python, let's try to rewrite in Python the Code 2.45 that sums the numbers from zero to 9 in the following Code 2.53:

```python
#Python script - for-operator example
count = 0
sum = 0
while (count < 10):
    sum = sum + count
    count = count +1
print(sum)
```

Code 2.53 Summatory example written in Python as in C++

We can not that similarly to the for-operator, the while ends as a semicolon and then two instructions are indented and belong to the while. The code is a crude translation in Python of the C++ one. It solves the same problem, but is it a correct approach in Python? The answer is not! Python is more flexible than C++, but it is an interpreted language. Therefore, if we program in Python as we did in C++, the code will be much less efficient. Let's try a different approach in the following Code 2.54:

```python
#Python script - for-operator example
data = range(10)
print(data.sum())
```

Code 2.54 Summatory in Python

In the example, we store a list of integer numbers from zero to 9 in the variable data and use the list's sum method to reach the same result. The loop seems to have disappeared in this case, but it is still present and will be made by sum(). The same exercise can be solved by using the Python built-in function sum, replacing the instruction *print(data.sum())* with *print(sum(data))*.

2.3.6 Applications

We report applications using what we learned about input-output operations, control structures, and loop operators.

2.3.6.1   Reading a File of Data Memorized in a Disordered Way

The first problem is reading a file where the data is stored in disorder. To create a context for this example, let's imagine having a device writing in a row of a file the number of charges produced by radiation on a detector in a minute, and working for several hours. This singular situation will produce a file of data where each row presents a random number of data points, because of the random nature of natural radioactivity. In this case, we cannot use a for-loop to read all the data because we do not know how much data we must read. The while-loop is the best solution, but not trivial because we must implement a method to stop the reading procedure. To manage these issues, the ifstream class is equipped with the eof() method (eof stands for end of file) that returns the logical value *true* if the object linked to the file reaches the end of the file and a false value in different cases. In the following code, the solution is achieved by using the eof() method:

```cpp
//the program tells a use
#include <iostream>
#include <string>
using namespace std;
int main(){
    ifstream readme("disorderedfile.dat");
    float value;
    int countme=0;
    while(!readme.eof()){
    readme>>value;
    countme++;
    }
    countme--;  //because the object must make at least a
    failed lecture
    return 0;
}
```

Code 2.55 Reading of a file with data stored in disordered way, calculation of average and root mean square of data sample

Another smart solution to the same problem is to use the property of the reading instruction to return a logic value when the reading succeeds or fails. In the following, the same Code 2.55 using this trick:

```cpp
//the program tells a use
#include <iostream>
#include <string>
using namespace std;
int main(){
    ifstream readme("disorderedfile.dat");
```

```cpp
7      float value;
8      int countme=0;
9      while(readme>>value){ //we read in the argument of the
       while-loop to use the returned logical value, true if the
        reading is successful, false otherwise
10     countme++;
11     }
12     cout<<" we read "<< countme<< "times"<< endl;
13     return 0;
14 }
```

Code 2.56 Read data with while-loop

Codes 2.55 and 2.56 solve the problem of counting the number of measurements stored in the file. Now we can proceed to analyze the data, for example, calculating the average and the root mean square of the data sample as follows:

```cpp
1  //the program tells a use
2  #include <iostream>
3  #include <string>
4  #include <cmath>//library to use the function pow()
5  using namespace std;
6  int main(){
7      ifstream readme("disorderedfile.dat");
8      float value;
9      int countme=0;
10     float sum =0; //initialize to zero the sum of all data
11      while(readme>>value){
12         sum+=value;
13         countme++;
14     }
15     cout<<"the average of read data is equal to "<<sum/
       countme<<endl;
16     float average = sum/countme;
17     //By reading again the file, we can estimate the root
       mean square
18     //as we did for the average, or we can define an array to
        store
19     //the data and to avoid reading a lot of time the same
       file in case
20     //of future analysis
21     float *array = new float[countme];
22     readme.close();
23     readme.open("disorderedfile.dat");
24     for(int i=0; i<countme; i++){
25     readme>>array[i];
26     }
27     //now we can estimate the root main squares
28
29     sum =0; //we recicle the variable
30     for(int i=0; i<countme; i++){
31     sum += pow(average - array[i],2);
32     }
33     float rms = sqrt(sum/countme);
```

```
34      cout<<"the root mean square is equal to "<<rms<<endl;
35      return 0;
36 }
```

Code 2.57 Reading of a file with data stored in disordered way

2.3.6.2 Reading a File of Data Memorized in a Disordered Way, in Python

In Python, the present exercise, reading a file containing data memorized in a disordered way, is easy concerning other languages, if we wish to read a file of real data, arranged into the file in a disordered way, it is enough to write the following script Code 2.58 using the Python built-in functions "open", "map" and "list" and the method of the string class "split":

```
1 data = list (map(float,open("data.dat").read().split()))
2 #just to verify that data contains a list of real values
3 for number in data:
4     print(number)
```

Code 2.58 Reading of a file with data stored in disordered way in Python

In only a row, the first one of the script, we opened the file, read it as a string, spitted the string in a list of strings, converted each string into a float value via map function, converted the object returned by map in a list and finally stored the results in the variable data. A long list of operations in a single instruction, of course, all functions perform the necessary loops and logical control on data. Still, the concept is that we do not have to invent the wheel every time, but to use Python at its best, we have to modify our programming approach, thinking about what Python offers before writing the code.

If we wanted to estimate the average value and the standard deviation of the data sample that we read and stored in the "data" list of the 2.58 code, we could translate the instructions of the C++ Code 2.57 into Python and this would be the wrong approach, even if it solves the problem. The best way is to use the Python functions and the properties of the Python list as reported in the Code 2.59:

```
1  import math # math library imported to use the sqrt function
2  data = list (map(float,open("data.dat").read().split()))
3  #just to verify that data contains a list of real values
4  for number in data:
5      print(number)
6  mean = sum(data)/len(data)
7  squaredeviation = []
8  for a in data:
9      squaredeviation.append((a-mean)**2)
10 rms = math.sqrt(sum(squaredeviation)/len(squaredeviation))
11 print("the mean values of read data is ", mean)
12 print("the standard deviation of data is ", rms)
```

Code 2.59 Mean value and standard deviation estimations in Python

The Code 2.59 has fewer instructions concerning the C++ Code 2.57, reflecting the high-level language of Python. The solution to the problem would have been even more straightforward. In line 6, we use the built-in Python functions *sum* and *len* to sum the data list, estimate the number of list elements, and calculate the average by the ratio of these two values. The loop in line 8 estimates the list of the squared deviation from the average value. Finally, in line 10, the standard deviation (or the root mean square) is estimated as the square root of the average of the square deviations. Please note that although Python code tends to be simpler and more compact, this doesn't imply that the complex tasks we've seen in C++ code are not being performed. The difference lies in Python's ability to delegate the handling of minor problems to built-in functions, allowing us to assemble code in a more modular, 'Lego-like' manner. To give an example of what we said, if we rewrite the Code 2.59 by using the powerful NumPy library we obtain a simpler Code 2.60:

```python
import numpy as np
data = np.loadtxt("data.dat")
print("the mean values of read data is ", data.mean())
print("the variance of data is ", data.var()
print("the Standard deviation of data is ", data.std())
```

Code 2.60 Mean value and standard deviation estimations via numpy

using just 4 instructions plus one to estimate the variance.

We will see the details, capabilities, and power of the NumPy library in Chap. 5.

2.3.6.3 Solving a System of Linear Equations by Gauss's Method Combined Use of Loops and Control Structures

The code presented in this paragraph solves the problem of finding the solutions of a linear system of 3 equations in 3 unknown variables with the Gauss method. The Code 2.61 can be easily generalized into systems of equations with any number of unknown variables. The program reads data from an external file and returns the system solutions.

```cpp
//Implementation of a code able to solve a linear system in
    three unknown variables with the Gauss triangularization
    method
#include <iostream>
#include <fstream>
using namespace std;
int main(){
/* indexes of coefficients of variables and known terms of
    the systems
    (0,0)x + (0,1)y + (0,2)z = (0,3)
    (1,0)x + (1,1)y + (1,2)z = (1,3)
    (2,0)x + (2,1)y + (2,2)z = (2,3)
the purpose of the code is to set the coefficients    (1,0),
    (2,0), e (2,1) to zero through linear combinations of the
    equations.
```

```cpp
11 It is easy to note that the zeros draw a triangle below the
       principal diagonal of the matrix of the known coefficient
        of the equation, which is why it is called the
       triangularization method.
12 The solution of the triangularized system becomes trivial,
       the solution of the last equation is immediate, and the
       others are derived in cascade by solving each equations
       from the last to the first.
13 */
14
15 // number of equations
16    int numrows = 3;
17 //number of coefficients
18    int numcoef    =numrows +1;
19    //matrix system
20    double coef[numrows][numcoef];
21    ifstream readme("data.dat");
22    if(!readme.is_open()){
23      cout<<"Sorry, I cannot open the file, please check!"<<
       endl;
24      return;
25    }
26 //reading loops
27    for(int i =0; i<numrows; i++){
28      for(int j=0; j<numcoef+1;j++){
29        readme>>coef[i][j];
30      }
31    }
32 //***************************!!
33
34 //Printing coefficients to check if read correctly!!
35    for(int i =0; i<numrows; i++){
36      for(int j=0; j<numcoef+1;j++){
37        cout<<coef[i][j]<<"   ";
38      }
39      cout<<endl;
40    }
41    cout<<endl;
42    cout<<endl;
43 //***************************!!
44
45 //Service variable
46    double denom, numen,temp;
47    int xx=1;//index to make row exchange
48 //***************************!!
49
50
51    for(int k =0; k<numrighe-1;k++){ //looping on the colouns
52      numen=coef[k][k]; // the element on the diagonal is the
       key of the process!
53 // if this coefficient is equal to zero, it does not allow us
       to perform the triangulation
```

```cpp
54 // so we proceed by searching for a row where the first
       element is different from zero in order
55 // to perform a simple swap
56     if(numen==0){
57 //this while-loop look for the first row useful for a swap
58        while(coef[k+xx][k]==0){
59     xx++;
60        }
61 //This loop perform the swap
62        for(int m =0; m<numrighe+1;m++){
63     temp = coef[k][m];
64     coef[k][m]=coef[k+xx][m];
65     coef[k+xx][m]=temp;
66        }
67 //  numen=coef[k][k];
68 //initialize xx to 1 in case of another control procedure via
        while-loop
69       xx=1;
70        }
71
72     for(int i =k+1; i<numrighe; i++){
73       denom=coef[i][k];
74       //If the element is zero in the triangle area is
       already zero, we skip the procedure
75       if(denom!=0){
76     for(int j=k; j<numrighe+1; j++){
77       coef[i][j] =  coef[k][j]- (coef[i][j]/denom)*numen;
78     }//looping on h
79        }
80     }//looping on i
81   }//looping on k
82
83   //Printing the final results
84   for(int i =0; i<numrighe; i++){
85     for(int j=0; j<numrighe+1;j++){
86       cout<<coef[i][j]<<"   ";
87     }
88     cout<<endl;
89
90   }
91
92 double sol[numrighe];
93
94 for(int i=numrighe-1; i>-1;i--){
95 sol[i]=coef[i][numrighe];
96
97 for(int j=i;j<numrighe-1;j++){
98 sol[i]-= sol[j+1]*coef[i][j+1];
99 }
100
101 sol[i]= sol[i]/coef[i][i];
102
103 }
```

```cpp
104  for(int i =0; i<numrighe;i++){
105  cout<<"soluzione "<<i+1<<" = "<<sol[i]<<endl;
106  }
107     return 0;
108  }
```

Code 2.61 Solving a system of linear equations by Gauss's method in C++

The code solves a linear system of three equations in three variables in the example. Still, it is straightforward to generalize to a system of many equations and variables.

2.3.6.4 Solving a System of Linear Equations with NumPy Methods

In the previous subsection, we reported the solution of a system of linear equations with Gauss's method in C++; the exercise was the occasion to force the reader to train the loop and logic structure. In Python, it is not necessary to reinvent the wheel every time. In the following Code 2.62 reports the solution to the problem by using a method of the Python class NumPy:

```python
1  #macro LinSys.py solving the system of linear equations whose
       coefficients are stored in the file test.dat
2  #data must be arranged with the variable coefficients and the
       known term of each equation in a single row and as many
       rows as the system equations.
3  import numpy as np
4  data = np.loadtxt("test.dat")
5  solutions = np.linalg.solve(data[:,0:len(data[:,0])],data[:,
       len(data[:,0])])
6  print(solutions)
```

Code 2.62 Solving a system of linear equations in Python

In Python, as usual, a few instructions correspond to many actions. To understand that it is enough to try to comment on the code. The first row imports the NumPy libraries and creates the alias in the word np, the second one imports the variable data the matrix of system equations coefficients stored in the file test.dat, and the third instruction uses the method solve of the method linalg to solve the system of linear equations, the last instruction print to terminal the solutions. Note that, according to the solve method's prototype, we passed it the matrix of variable coefficients and the array of known terms, but we performed that by properly slicing the data stored in the NumPy array.

2.3.6.5 Search for Absolute Maximum and Minimum

Another interesting example is the search for the absolute maximum and minimum in a data sample. We have already seen how to read a file and store data in an array, so the strategy will be to compare each array element with the right condition to determine the maximum or minimum value. In the following, we report just the piece of Code 2.63 solving the task:

```
1  //we declare two variables to store the absolute maximum and
        minimum
2  //initializing to the first element of the array
3  float max=misure[0];
4  float min=misure[0];
5  for(i=0; i<DimArray; i++){
6    if(data[i]>max){
7      max=data[i];}
8    else if(data[i]<min){
9    min=data[i];}
10 }
11 cout<<"that minimum value is: "<<min<<endl;
12 cout<<"that maximum value is: "<<max<<endl;
```

Code 2.63 Code looking for absolute maximum and minimum values in a collection of numebers in C++

We invite the reader to consider why the max and min variables are initialized with a value from the array and what could happen if they were initialized arbitrarily. In both cases, the code may run without errors, but in the second scenario, it might fail to correctly identify the maximum or minimum value if the initial value doesn't belong to the data set. Why does this happen?

2.3.6.6   Search for Absolute Maximum and Minimum with Python

We solve the problem with two approaches. The first one uses Python functions: we read data and store it in a list, and we use the Python built-in functions *min* and *max* on the list.

```
1  data = list(map(float, open("facilino.dat").read().split()))
2  print("the minimum is ",min(data), "  the maximum is ", max(
        data))
```

The second method solves this simple problem using the numpy library.

```
1  import numpy as np
2  data = np.loadtxt("facilino.dat")
3  print("the minimum is ",data.min(), "  the maximum is ", data
        .max())
4  #numpy furnishes other useful methods:
5  print("the mean value is ",data.mean(), "  the variance is  "
        , data.var())
```

NumPy is a powerful tool for problems related to numbers. The examples are just for the first practice; in the book, we'll report more complex uses.

2.3.6.7 Histogram Computation in C++

Before writing a program, you must fully understand the problem and think of the best computation strategy.

So, first of all, let's try to understand what a histogram is. A histogram is an "ordered" way of representing data. It is essential in the case of repeated measurements of the same physical quantity (single-dimensional histogram) or of multiple quantities concerning each other (multidimensional histogram), and it allows us to read the statistical information contained in the data quickly.

A practical approach to building a histogram involves identifying the absolute maximum and minimum values within the dataset. These values define the total range of variation for the histogram. This range is then divided into a chosen number of equal-width sub-intervals, commonly referred to as 'bins'. Finally, the number of data points that fall within that interval is counted for each bin.

The code shown in Listing 2.64 performs this task by printing, in one column, the average value of each sub-interval, and in a second column, the absolute frequency of the counts of the bins.

```cpp
//Histogram computation
int rebinning=0;
do{ // allows to repeat and to change the bin intervals
  int nint;
  cout<<"\nLet me know the number of bins: \n";
  cin>>nint;
  float l=(max-min)/(float)nint; //Bin width estimation
  cout<<"Bin width: "<<l<<endl;
  int bin[nint]; //defines an array of counters
  for(i=0;i<nint;i++){
     bin[i]=0; //initializes each counter to zero
     }
  for(i=0;i<nint;i++){
    for(int v=0; v<j; v++){
    if((misure[v]>=min+i*l)&&(misure[v]<=min+(i+1)*l)){
//Counts the frequency of numbers in each i interval
    bin[i]++;
     }
   }
  }

//cout << scientific;
  cout << fixed; //fixes the alignment
  for(int n=0; n<nint; n++){
   cout<<min+l/2+n*l<<"\t"<<bin[n]<<endl;
  }
  cout<<"\n Do you want to change the number of bins? (if yes
      digit 1):\n";
  cin>>rebinning;
}while(rebinning==1);
```

Code 2.64 Histogram computation in procedural mode in C++

The output of the code is:

```
-4.049477    21
-2.526830    154
-1.004183    1038
0.518463     4618
2.041110     13894
3.563756     27787
5.086403     36796
6.609050     31854
8.131697     18555
9.654344     7240
11.176990    1821
12.699636    329
14.222283    40
15.744930    3
17.267576    0
```

What we observe is a numerical, rather than graphical, representation. As mentioned earlier, the table provides a summary of static information. Even at a glance, it's evident that the highest frequency occurs within the interval between 5.086403 and 6.609050. This suggests that the distribution's average value lies within this range. Indeed, in the direct calculation of the mean shown in the previous examples, the result was 5.35344.

2.3.7 Histogram Computation with Python

We repeat the previous section's exercise to show how the approach changes when we solve the same problem in a high-level language like Python. The rule of the game is to read data from a file, determine the maximum and the minimum value, and compute the frequency of data in a desired number of intervals (bins) between the maximum and the minimum values of data. We will check if some Python tools can solve the tasks above. We see that NumPy, which we learn more about later in the book, can read data, find the maximum and minimum value in a data collection, and compute a histogram. So, the most convenient way is to check the right syntax to use NumPy methods and proceed to solve the problem. The following Code 2.65 reports a simple Python script able to compute the histogram:

```python
import numpy as np
data = np.loadtxt("test.dat")
histo = np.histogram(data,bins=50,range = (data.max(),data.
    min()))
print("first array contains the histogram absolute frequency"
    )
```

```
5 print(data[0])
6 print("the second array contains the edges of histogram bins"
      )
7 print(data[1])
```

Code 2.65 Histogram computation with Numpy

The script loads the data numbers in a NumPy array called data, then the method histogram computes the histogram taken as input from the array containing data, the number of bins, and the absolute edges of histograms computed by NumPy with the methods max() and min(). The method histogram returns a tuple containing two arrays, the first one stores the absolute frequency of the histogram, while the second one stores the edges of the histogram bins. If you wish to use the information for a graphical representation of the histogram, remember that the second array does not store the center of bins, but the bin edges. The second array is of dimension equal to the first array plus one.

2.3.8 Functions

In programming, functions provide the primary means of making a program modular. They are designed to solve specific problems, accepting data as input, processing it, returning values, or modifying the variables they receive. By delegating specific tasks to functions, complex problems can be broken down into smaller, manageable parts, enhancing code modularity and readability.

2.3.8.1 C++

In C/C++, every program and subprogram is defined as a function. Even the main program is a function, named *main*, because it represents the entry point and must be present in every compiled C++ program. All functions, including *main*, follow the same structure: specify the return type, function name, and input parameters within parentheses. The body of the function is enclosed in curly braces.

The following code is a simple function implementation that returns the sum of two integers provided to the function.

```
int sum(int a, int b){
return a+b;
}
```

Now let's see how functions are implemented within a program, how they are called, and what rules must be followed. One rule that does not allow exceptions is that every function, like variables, must be declared before executing the *main* function. The declaration of the function is called a prototype. The declaration may or may not coincide with the full implementation. The prototype of a function consists of its return type, the function name, and the parameters it takes. The names of variables used within function implementations are entirely arbitrary, and in general, the variable names used in the *main* function do not need to match those in other functions. Variables declared inside a function are considered local and are only accessible within that function. To create global variables that are accessible to all functions, including the *main* function, they must be declared before any function definitions, typically after the inclusion of libraries. Below is an example showing a function prototype and its implementation defined before the *main* function 2.66.

```cpp
#include <iostream>
using namespace std;
int sum(int a, int b){
return a+b;
}
int main(){
int var1,var2;
var1=10;
var2=20;
cout<<somma(var1,var2)<<endl;
return 1;
}
```

Code 2.66 Function prototype declaration and implementation before the main function

Another example of function prototyping, implemented after the main function, is reported in Code 2.67.

```cpp
#include <iostream>
using namespace std;

int somma(int a, int b);

int main(){
int pino,pina;
pino =10;
pina=20;
cout<<somma(pino,pina)<<endl;
return 1;
}
int somma(int a, int b){
return a+b;
}
```

Code 2.67 Function prototype declared before the main function and implementation after

An example where the auxiliary function "sum" is written in a header file "sum.h" is reported in Code 2.68:

```cpp
#include <iostream>
#include "somma.h"
using namespace std;
int main(){
int pino,pina;
```

```
 6  pino =10;
 7  pina=20;
 8  cout<<somma(pino,pina)<<endl;
 9  return 0;
10  }
```

Code 2.68 Function prototyping and implemented in a header file

The header file "sum.h", which in this example resides in the same folder as the program that loads it through the #include directive, contains the prototype and implementation of the function:

```
// file somma.h
int somma(int a, int b){
return a+b;
}
```

Passing Values to Functions by Value or Reference in C++

In the previous section's examples, we implicitly used the technique of passing arguments by value. When passing variables from one function (such as the main function) to another, avoiding unintentional changes to the original data is crucial. To prevent this, C/C++ uses pass-by-value as the default method. This means that the called function receives a copy of each argument and operates on these copies, leaving the original variables unchanged.

However, if we want the function to modify the variables passed directly, we can use a different approach, known as passing value *by reference*. In this method, we do not pass the variable to the function but rather the address where the variable is stored. The function will copy this information as before, but can act on the variables through the memory addresses. A variable able to store a memory address is called a "pointer". The *pointer* variables are the keys for passing values by reference.

Let's write an example

```
 1  // Function prototype able to read addresses
 2  // through pointer variables
 3  void swap(double *first, double *second);
 4
 5  int main(){
 6    double a = 5.4;
 7    double b = 10.2;
 8    cout << "Value of a: " << a << endl;
 9    cout << "Value of b: " << b << endl;
10    swap(&a, &b);
11    cout << "New value of a: " << a << endl;
12    cout << "New value of b: " << b << endl;
13    return 0;
14  }
15
16  void swap(double *first, double *second){
17    double temp;
```

```cpp
18    temp = *first; // accessing the content of the pointer
19    *first = *second;
20    *second = temp;
21    return 0;
22 }
```

Code 2.69 Correct implementation of a swap function in C++, passing data by reference

We implemented the classic swap function in the code example referenced as 2.69. We passed the addresses of variables a and b to the swap function, storing them in two pointer variables. The function then uses these pointers to access and modify the values at the corresponding memory addresses. To clarify, the variables first and second hold the memory addresses of a and b, respectively. We can access and modify the contents of these memory locations by using the dereference operator (*).

If we try to repeat the same exercise using the method of passing by value to the function, we can rewrite the Code 2.69 in the following Code 2.70:

```cpp
1  #include <iostream>
2  // Function prototype capable of reading addresses
3  // through pointer variables
4  void swap(double first, double second);
5
6  int main(){
7    double a = 5.4;
8    double b = 10.2;
9    cout << "Value of a: " << a << endl;
10   cout << "Value of b: " << b << endl;
11   swap(a, b);
12   cout << "New value of a: " << a << endl;
13   cout << "New value of b: " << b << endl;
14   return 0;
15 }
16
17 void swap(double first, double second){
18   double temp;
19   temp = first;
20   first = second;
21   second = temp;
22   return;
23 }
```

Code 2.70 Wrong implementation of a swap function in C++, passing data by value

And we wonder what would happen if we executed this code. The answer is that the old and new values would coincide, the swap function would act on copies, and the original variables would remain as they were.

If we want a function to modify the original variables passed to it, we must pass them by reference. This means we don't pass the variables themselves, but rather their memory addresses. Although the function receives a copy of these addresses, it can use them to access and modify the original variables directly. The function must be implemented using pointer variables, which operate on the values stored at the given memory locations.

The syntax and the use of pointer variables are explained in the following Code 2.71:

```cpp
1  // Pointer variable example
2  #include <iostream>
3  using namespace std;
4
5  int main(){
6    double normalvariable = 10;
7    //Declaration of pointer variable
8    //kind  *nameofvariable;
9    double *pointervariable;//the symbol * preceded the
       variable
10   pointervariable = &nomaralvariable;//the symbol & allows us
        to access to the memory address where the information of
        the normalvariable is stored.
11   cout<<"what is stored in the memory address of the normal
        variable by using the pointer"<<*pointervariable; //Only
        after the definition, the symbol * allows to access the
        value pointed by the memory address stored in the
        pointervariabel
12   return 0;
13 }
```

Code 2.71 Use and meaning of pointer variables

The variable declared on line 7 of Code 2.71 is not a pointer; however, like all variables, its value is stored at a specific memory address in the computer. On the other hand, the variable declared on line 10 is a pointer designed to hold the address of a double-precision floating-point value (double). Initially, this pointer is uninitialized, as pointer variables do not contain valid addresses until explicitly assigned. In line 10, we assign it the memory address of the variable from line 7 using the address-of operator (&). Finally, line 12 shows how we can access the original variable's value through the pointer using the dereference operator (*).

Recursion

Recursion is a fundamental property of the function in computer programming and consists of the ability of the function to call itself. It is very useful in all codes using the recursive search. A usual example of recursion consists of the implementation of factorial, reported in Code 2.72:

```cpp
1  // Factorial computation
2  #include <iostream>
3  using namespace std;
4  long factorial (long value)
5  {
6    if (value > 1){
7      return (value * fattoriale (value-1));
8    }
9    else{
```

```
10      return 1;
11    }
12 }
13
14 int main ()
15 {
16    long data = 9;
17    cout << data << "! = " << factorial (data);
18    return 0;
19 }
```

Code 2.72 Recursion function example

2.3.8.2 Functions in Python

In Python, functions are defined using the "def" keyword followed by the function
name, a set of round parentheses enclosing any arguments, and a colon. The body
of the function is indented and contains the statements that define what the function
does. The "return" statement is used within the indented body to return a value or
exit the function. Let's start with a simple Python script 2.73:

```
1 Def printme(x):
2     print("the number passed to me is ", x)
3     return
4 #to test the script
5 printme(3.14)
```

Code 2.73 Implementation of a simple function printing a given number in Python

Functions in Python can be defined anywhere within the script. However, it's important to ensure that the function is defined before it is called. If you try to call a function
before it is defined, Python will raise a NameError because it won't recognize the
function. If we try to violate this recommendation, for example by inventing the
calling to the function in the previous Code 2.73 as follows 2.74:

```
1 #to test the script
2 printme(3.14)
3
4 Def printme(x):
5     print("the number passed to me is ", x)
6     return
```

Code 2.74 A wrong way to implement function in Python

Python will not execute the script and write :

```
Traceback (most recent call last):
  File "testfunc.py", line 1, in <module>
    printme(3.14)
NameError: name 'printme' is not defined
```

As an exercise, we rewrite the function sumtwonumbers implemented in C++ in the previous sections as follows in the Code 2.75:

```python
def sumtwonumbers(first, second):
    return first + second
print(" 10 + 2 is equal to ", sumtwonumbers(10,2))
```

Code 2.75 Example of a function summing two numbers in Python

The value passed to the function is implicit in the kind of variables passed. If the variable is not an address, the pass is *by value*, if it is an address, by reference. For example, the following attempt to implement a swapping function fails; the script 2.76 is synthetically correct but doesn't do what we wish:

```python
def swap(a,b):
    c=a
    a=b
    b=c
    return
x=10
y=20
print(x," ",y)
swap(x,y)
print(x," ",y)
```

Code 2.76 Example of a swapping function in Python that does not working

The execution of the script writes on the screen ten and twenty in the same sequence, two times:

```
10    20
10    20
```

What happens if we change the variable into a list? Let's try to implement it in the following Code 2.77:

```python
def swap(a,b):
    c=a
    a=b
    b=c
    return
x=[]
y=[]
x.append(10)
y.append(20)
print(x," ",y)
swap(x,y)
print(x," ",y)
```

Code 2.77 Modified swapping function implemented in Code 2.76 that still does not working

The results of the improved code are:

```
[10]    [20]
[10]    [20]
```

What wrong? Why does the code not do what we expected? Because we passed two addresses, but in the function, we do not act on the value pointed to by the address variables. Correcting the Code 2.78 in the following way:

```python
def swap(a,b):
    c=a[0]
    a[0]=b[0]
    b[0]=c
    return
x=[]
y=[]
x.append(10)
y.append(20)
print(x," ",y)
swap(x,y)
print(x," ",y)
```

Code 2.78 Example of a swapping function in Python working properly

finally we obtain:

```
[10]    [20]
[20]    [10]
```

Built-in Functions in Python

Python offers several built-in functions. The complete list is available on the official website of the Python language at the following web-page. Many built-in functions are already used in this chapter: the functions type, print, input, range, len, min, max, and sum.

2.4   Exercises

2.1 Write a code displaying a message on the screen.

2.2 Write code capable of asking a user for a number and printing it on the screen.

2.3 Write code capable of asking a user for two real numbers and printing their sum on the screen.

2.4 Write code capable of asking a user for two real numbers and asking which of the 4 basic operations they want to perform, then printing the result on the screen. Make the code iterative, capable of repeating the operations unless the user specifies otherwise.

2.5 Write code capable of asking for an integer and indicating whether it is even or odd.

2.6 Ask a user to provide 10 numbers and store them in an array. Calculate the sum of the provided numbers and print the final result on the screen.

2.7 Write a program filling the first column of a matrix with even numbers from 0 to 100, the second column with odd numbers, and the third column with the sum of the corresponding values from the first two columns.

2.8 Write code that reads the file "data.dat," prints all the values on the screen, and calculates the mean and standard deviation.

2.9 Write code that counts the rows and columns in the file "data.dat".

2.10 Write code to find the absolute maximum and minimum of the fifth column of the file data.dat.

Chapter 3
Why Object-Oriented Programming?

3.1 Introduction

Object-Oriented Programming (OOP) is a programming paradigm that uses the concept of "objects" to design and structure software. Some of the most important aspects of OOP are the following:

- **Classes and Objects**:

  - Class: A blueprint for creating objects. It defines a set of attributes and methods that the created objects will have.
  - Object: An instance of a class. It represents a specific entity with attributes and behaviours defined by its class.

- **Encapsulation**, the practice of bundling data (attributes) and methods (functions) that operate on the data into a single unit, or class. It restricts direct access to some of the object's components, which can prevent the accidental modification of data.
- **Inheritance** allows a new class to inherit attributes and methods from an existing class. This promotes code reusability and establishes a natural hierarchy between classes.
- **Polymorphism** allows objects of different classes to be treated as objects of a common superclass. It enables a single interface to represent different underlying forms (data types)
- **Abstraction**, which involves hiding the complex implementation details of a system and exposing only the necessary and relevant parts. It helps reduce complexity and allows the programmer to focus on interactions at a higher level.

OOP is widely used in many programming languages, including C++, Java, Python, and more. It helps in creating modular, reusable, and maintainable code, making it a fundamental concept in software development.

© The Author(s), under exclusive license to Springer Nature Switzerland AG 2026   75
S. Vasi et al., *Open Source Tools for Physics Data Analysis*, Undergraduate Texts
in Physics, https://doi.org/10.1007/978-3-032-00721-6_3

3.2   Class

In programming, the concept of a class extends the idea of a struct in C and forms the foundation of object-oriented programming (OOP). Classes provide powerful features such as encapsulation, inheritance, and polymorphism—each complex enough to warrant its dedicated study. However, this text aims to introduce the basic implementation and practical use of classes in data analysis. Readers are encouraged to consult specialized programming literature for a more comprehensive understanding of OOP concepts.

A class is a data structure that combines data (attributes) and functions (methods) operating on that data. The core principle behind implementing a class is to design a structure that addresses a given problem as generally and flexibly as possible, providing the necessary tools for managing the data and its associated operations.

A useful—though simplified—analogy is to think of a class as a biological cell: just as a cell takes in substances, processes them, and produces results, a class receives input, processes it through its methods, and returns output. The class serves as the conceptual blueprint for solving or managing a problem, while the object—an instance of the class—is what performs the operations. Functions defined within a class are referred to as its methods.

In C++, a class is declared using the class keyword, followed by the class name. Its contents are enclosed in curly braces. By default, all members (data and methods) are private, which means they are accessible only from within the class. To make them accessible from outside, you must explicitly declare them as public using the *public:* keyword. You can modify the access level of class members at any point during the class definition using the private, public, or protected specifiers.

Below is an example of a simple class that performs basic mathematical operations on two real numbers, as shown in Code 5.1.

```
#include <iostream> using namespace std; class operation{
//private by default
float first, second;
    public://the subsequent instructions are public
    operation(){ //first constructor
        return;
    }
    operation(double a, double b){ //second constructor
        first = a;
        second = b;
        return;
    }
    void changeFS(float a, float b){
        first = a;
        second = b;
```

```cpp
16          return;
17      }
18      void changeF(float a){
19          first =a;
20          return;
21      }
22      void changeS(float a){
23          second =a;
24          return;
25      }
26      float Sum(){
27          return first + second;
28      }
29      float FminusS(){
30          return first - second;
31      }
32      float SminusF(){
33          return second - first;
34      }
35      float product(){
36          return first*secondo;
37      }
38      float FdividedbyS(){
39          return first/second;
40      }
41      float SdividedbyF(){
42          return second/first;
43          ~operation(){
44              cout<<"hello, I'm killing your object"<<endl;//
    Comment just as a joke! Normally no comment in the
    distructor
45          return;
46          };
47      }
48 };//note the semi column at the end of the implementation
49 int main(){
50     //first example of the use of the class
51     operation objcalc1; //objcalc is an object of operation
     class
52     float numb1, numb2;
53     court<<"give me two numbers"<<endl;
54     cin>>numb1>>numb2;
55     objcalc1.changeFS(numb1, numb2);//operator point "."
    give access to methods!
56     cout<<numb1<<" + "<<numb2<<" = "<<objcalc1.sum()<<endl;
57     //second example of the use of the class
58     operation objcalc2(numb1,numb2);//we using the second
    constructor
59     cout<<numb1<<" - "<<numb2<<" = "<<objcalc2.FminusS()<<
    endl;
60     //third example of the use of the class
61     operation *objpointercalc = new operation(numb1,numb2);
62     cout<<numb1<<" / "<<numb2<<" = "<<objpointercalc->
    FdivededbyS()<<endl;
63     //in the last case, the object pointer variable
```

```
64      //access to method by operator arrow "->" (minus +
        major symbols)
65      return 0;
66  }
```

Code 3.1 Example of implementation of a class managing the 4 fundamental mathematical operation

Once an object of a class has been defined, the way to access the class members is through the dot operator or the arrow operator (formed by combining the minus and greater-than symbols). The dot operator is used when the object is a regular variable, while the arrow operator is used when the object is a pointer.

Let's comment on the Code 5.1, the first question is what private and public mean? The variables "first" and "second" of the class are implicitly defined as private, which means that the only way to modify their values is through the method of the class itself. If you try, in the main program, to change the value of this variable directly, you get an error. See the following code as an Example 3.2:

```
1  //  ....//

2

3  int  main(){

4

5      operation objcalc1;
6      objcalc1.first = 12.34;  //this is a mistake!!!
        the variable first is private
7      objcalc1.changeF(12.34);  //this is a correct way
        to change the value!
8      return 0;
9  }
```

Code 3.2 Example on how to change the value of a private variable in a C++ class

However, if the variables "first" and "second" had been declared public, the instruction $objcalc1.first = 12.34;$ would have been correct. Class data should be kept private for security and code stability, especially in large software projects.

All functions defined within a class are called its methods. Methods that share the same name as the class are known as constructors. The destructor, a special method used to destroy the object and free allocated memory, is identified by the class name preceded by a tilde ( ) symbol. Constructors do not specify a return type because, in C++, they cannot return values. Multiple constructors can be defined within a class, differing in the number or types of parameters. This allows for the selection of the appropriate constructor when creating an object. For example, in the Code 5.1, the statement operation objcalc1; invokes the default (empty) constructor. In contrast, operation objcalc2(numb1, numb2); calls a second constructor that initializes internal variables with the values numb1 and numb2. This second constructor acts as a shortcut, simultaneously creating the object and setting its internal state. In other words, operation objcalc2(numb1, numb2); is equivalent to writing:

```
1 operation objcalc1;
2 objcalc1.changeFS(numb1, numb2);
```

The use of the class in the code reported in 5.1 shows us that the class is an abstract definition of rules useful to manage a problem, But operatively, the use of the class passes through the object, the variable of the kind defined by the class.

3.2.1 *Making Practice with Class*

The writing of a class should be guided by the idea that it should help solve problems in the most general way possible, to avoid the proliferation of classes that perform similar functions and unnecessary code repetition. Even when writing code, one should focus on resource efficiency.

Perhaps it's not the most straightforward example, but let's try to write a class that produces several outputs based on two input values. In Example 3.3, we aim to define a class named RegularPolygon, which takes the base and height as input parameters. Upon request, this class will be able to return the perimeter and area of various geometrical shapes, specifically a triangle, square, rectangle, or pentagon. We'll assume the usual geometric interpretations: for the triangle and rectangle, base and height have their standard meanings; for the square, the base will be used as the side length; and for the pentagon, the height will represent the apothem.

```
1  class RegularPolygon {
2    double base;    //Private by default
3    double height;
4    public:
5    //Prototyping for external implementation
6    void SetBaseHeight(double,double);
7    double TriangleAreas ();
8    //Implementation inside the class declaration
9    double TrianglePerimeter(){
10         return 3*base;
11   }
12   double RettangleAreas(){
13         return base*height;
14   }
15   double RettanglePerimeter(){
16         return 2*(base+height);
17   }
18   double SquareAreas ();
19   double SquarePerimeter ();
20   double PentagonAreas ();
21   double PentagonPerimeter();
22 };
23
24 //For example, external implementation of methods
25 //RegularPolygon
26 void RegularPolygon::SetBaseeheight(double a, double
       b){
```

```
27    base=a;
28    height=b;
29    return;
30    }
31 double RegularPolygon::TriangleAreas(){
32        return base*height/2;
33 }
34 double RegularPolygon::SquarePerimeter(){
35        return base*4;
36 }
37 double RegularPolygon::SquareAreas(){
38        return base*base;
39 }
40 double RegularPolygon::PentagonPerimeter(){
41        return base*5;
42 }
43 double RegularPolygon::PentagonAreas(){
44        return (base*height/2)*5;
45 }
46
```

Code 3.3 Example of internal and external implementation of methods of a C++ class

In the class shown above in Code 3.3, you may notice that some methods are implemented directly within the class definition. In contrast, others are defined externally using the scope resolution operator (::) to link the method to its class. This isn't meant to promote inconsistency, but to demonstrate that both approaches are valid in C++.

In typical practice, the class declaration is placed in a header file with a .h extension, while the method implementations are written in a separate .cpp source file.

Another critical point is that the base and height variables are declared without access modifiers, which means they are private by default. As with square brackets, it's crucial to understand this default behavior but not rely on it excessively. It's considered good practice to specify the access level of class members and methods explicitly.

Access levels are defined using the keywords public, private, and protected, followed by a colon. These modifiers apply to all subsequent members until another access specifier appears.

In the example below, variables A, B, and D are public, while variable C is private:

```
class pippo{
public:
double A,B;
private:
int C;
public:
char E;
};
```

3.2.2 Inheritance: An Example

In the first Example 3.4, we implement a simple class to define the basic data required
to describe the shape of a polygon. We then create a Rectangle class that inherits from
this base class, gaining access to its properties. This kind of example is commonly
found in many C++ tutorials available online.

```cpp
#include <iostream>

using namespace std;

class shape{
protected:
double base, height;
public:
shape();
shape(double, double);   //constructor
//~shape();        //destructor
void SetBaseHeight(double a, double b){
   base=a;
   height=b;
   return;
}
double GetHeight(){
   return height;
}
};

//Constructor external implementation
shape::shape(){
cout<<"Just created a shape class object"<<endl;
}
shape::shape(double p, double m){
    base=p;
    height=m;
}

//implementazione distruttore
shape::~shape(){
    cout<<"\n destroyed"<<endl;
}

//Derivate class
class rectangle:public shape{
public:
rectangle(double,double);    //Constructor
~rectangle();                //destructor
double area(){
   return base*height;
   }
};

//Constructor implementation
```

```cpp
47  rectangle::rectangle(double a, double b){
48    base=a;
49    height=b;
50  };
51
52  //implementazione distruttore
53  rectangle::~rectangle(){
54    cout<<"destroyed"<<endl;
55  }
56
57  int main(){
58    cout<<"this program estimate the areas of a
          rettangle\n";
59    double a, b;
60    cout<<"give me base and height of rectangle: "<<
          endl;
61    cin>>a>>b;
62    shape pippo(a,b);
63    cout<<"Testing the GetHeight method: "<<pippo.
          GetHeight()<<endl;
64    rectangle pina(a,b);
65    cout<<"Areas is : "<<pina.area()<<endl;
66    return 0;
67  }
```

Code 3.4 Example of Inheritance of the C++ class

3.2.3 Examples

In this Example 3.5, we will repeat the previous exercise but in a case familiar
to physics problems. We will implement a general class called Vector, which will
provide methods to load the necessary information to characterize a vector in three-
dimensional space (its components). Then, we will create a derived class, VectorOp-
erations, which will perform operations on two vectors and use the data and methods
inherited from the parent class to provide information about the resulting vector from
the operations performed.

```cpp
1      #include <iostream>
2  #include <cmath>
3
4  using namespace std;
5
6  class vector {
7  protected:
8      double x, y, z;
9  public:
10     vector();
11     vector(double a, double b, double c) {
12         x = a;
13         y = b;
14         z = b;
15     }
```

```cpp
    void SetX(double pippolo) {
        x = pippolo;
    }
    void SetY(double paulo) {
        y = paulo;
    }
    void SetZ(double peulo) {
        z = peulo;
    }
    void SetXYZ(double a, double b, double c) {
        x = a;
        y = b;
        z = b;
    }
    double GetX() {
        return x;
    }
    double GetY() {
        return y;
    }
    double GetZ() {
        return z;
    }

    double GetTheta() {
        return atan2(sqrt(pow(x, 2) + pow(y, 2)), z)
    * 180 / 3.14159265;
    }

    double GetPhi() {
        double phi = atan2(y, x) * 180 / 3.14159265;
        if (phi < 0) {
            return phi + 360.;
        }
        else {
            return phi;
        }
    }

    ~vector();

};

vector::vector() {
    cout << "I have constructed an object of type
    vector" << endl;
}

vector::~vector() {
    cout << "Hello Dear!!!!" << endl;
    cout << "I just destroyed an object of type
    vector" << endl;
```

```cpp
66      cout << "Sincerely, the Destructor" << endl;
67  }
68
69  class Operations: public vector {
70
71  private:
72      double x1, y1, z1, x2, y2, z2;
73
74  public:
75
76      Operations();
77      Operations(double a, double b, double c, double
    aa, double bb, double cc) {
78          x1 = a;
79          y1 = b;
80          z1 = c;
81          x2 = a;
82          y2 = b;
83          z2 = c;
84      }
85
86      void SetVector1(double a, double b, double c) {
87          x1 = a;
88          y1 = b;
89          z1 = c;
90      }
91      void SetVector2(double aa, double bb, double cc)
        {
92          x2 = aa;
93          y2 = bb;
94          z2 = cc;
95      }
96      void SetVectors(double a, double b, double c,
    double aa, double bb, double cc) {
97          x1 = a;
98          y1 = b;
99          z1 = c;
100         x2 = a;
101         y2 = b;
102         z2 = c;
103     }
104
105     void Sum() {
106         x = x1 + x2;
107         y = y1 + y2;
108         z = z1 + z2;
109     }
110 };
111
112 Operations::Operations() {
113     cout << "Hello" << endl;
114 }
115
```

```cpp
116  int main() {
117
118      double n, m, s;
119
120      cin >> n >> m >> s;
121
122      vector pino(n, m, s);
123      cin >> n >> m >> s;
124
125      vector pina(n, m, s);
126
127      Operations test(n, m, n, m, n, s);
128      test.Sum();
129      cout << test.GetTheta() << endl;
130
131      n = 10;
132
133      return 0;
134  }
```

Code 3.5 Implementation of a C++ class managing the problem of a 3D vector

The following Example 3.6 is an implemented code able to compute a histogram.

```cpp
1        #include <iostream>
2  #include <cstdlib>
3  #include <fstream>
4  #include <cmath>
5
6  using namespace std;
7
8  class data {
9      private:
10          int index;
11          double *vector;
12          int *histogram;
13          double max, min;
14          double mean, stddev, meanDev;
15          string filename;
16
17      public:
18          data(string);                      // Constructor
     prototype (loads data from an external file)
19          data(double*, int);                // Constructor
     prototype (loads data from an internal array)
20          int loadFromFile(string name) // Method to
     load data from a file
21          {
22                  index = 0;
23                  double temp;
24                  ifstream readFile(name.c_str());
25                  if (!readFile.good()) {
26                          cout << "Error! File not found." <<
     endl;
27                  }
```

```cpp
        while(!readFile.eof()) {
            readFile >> temp;
            if(!readFile.eof()) {
                index++;
            }
        }
        vector = (double *)malloc(index * sizeof
(double));
        readFile.close();
        readFile.open(name.c_str());
        for(int i = 0; i < index; i++) {
            readFile >> *(vector + i);
        }
        readFile.close();
    }
    double show(int element)        // Method to
display an element at the nth position in the
array
        {
        element--;                      // Since the
first index of the array is 0, the index of the
desired element is equal to the desired position
minus one
        if (element < index) {
            return *(vector + element);
        } else {
            cout << "Error! The chosen index
exceeds the maximum index." << endl;
            cout << "\n\nCritical error #";   //
Just for dramatic effect and since the method
must return something, we give meaning to that
something
            return 0;
        }
    }
    double getMax()                      // Method to
find the maximum value
        {
        max = *vector;
        for (int i = 0; i < index; i++) {
            if (*(vector + i) >= max) {
                max = *(vector + i);
            }
        }
        return max;
    }
    double getMin()                      // Method to
find the minimum value
        {
        min = *vector;
        for (int i = 0; i < index; i++) {
            if (*(vector + i) <= min) {
                min = *(vector + i);
```

```cpp
            }
        }
        return min;
    }
    double getMean()                         // Method to
calculate the mean value
    {
        mean = 0;
        for (int i = 0; i < index; i++) {
            mean += *(vector + i);
        }
        mean /= index;
        return mean;
    }
    double getStdDev()                       // Method to
calculate the standard deviation
    {
        double sumSqDiff = 0;
        getMean();                           // Calculate
the mean
        for (int i = 0; i < index; i++) {
            sumSqDiff += pow(*(vector + i) -
mean, 2);
        }
        stddev = sqrt(sumSqDiff) / sqrt((double)
index);
        return stddev;
    }
    double getMeanDev()                      // Method to
calculate the standard error of the mean
    {
        getStdDev();
        meanDev = stddev / sqrt(index);
        return meanDev;
    }
    int getSize()                            // Method to
return the number of data points in the file
    {
        return index;
    }
    void createHistogram(int subint)  // Method
to create a histogram with 'subint' intervals
    {
        histogram = new int[subint];
        double range = (max - min) / subint;
        getMin();                            // Calculate the
minimum value
        getMax();                            // Calculate the
maximum value
        for (int i = 0; i < subint; i++) {   //
Reset the bins
            *(histogram + i) = 0;
        }
```

```cpp
            for (int i = 0; i < subint; i++) {
                for (int j = 0; j < index; j++) {
                    if ((*(vector + j) >= min + i *
range) && (*(vector + j) < min + (i + 1) * range)
) {
                        *(histogram + i) = *(
histogram + i) + 1;   // Increment bin i+1 by one
                    }
                }
            }
            *(histogram + subint - 1) += 1;   //
Increment the last bin by one since the maximum
was excluded
        }
        double showHistogram(int position)  //
Method to display a bin
        {
            return *(histogram + position - 1);
        }
};

data::data(double *array, int index)  // Constructor
    with input of an array and its size
{
    this->index = index;
    vector = new double[index];
    for (int i = 0; i < index; i++) {  // Load the
array into the class
        *(vector + i) = array[i];
    }
}

data::data(string name)
{
    index = 0;
    double temp;
    ifstream readFile(name.c_str());
    if (!readFile.good()) {
        cout << "Error! File not found." << endl;
    }
    while(!readFile.eof()) {
        readFile >> temp;
        if(!readFile.eof()) {
            index++;
        }
    }
    vector = (double *)malloc(index * sizeof(double)
);
    readFile.close();
    readFile.open(name.c_str());
    for (int i = 0; i < index; i++) {
        readFile >> *(vector + i);
    }
```

```cpp
155        readFile.close();
156 }
157
158 int main()
159 {
160     int intervals;
161     char response;
162     string filename;
163     cout << "Which file would you like to load data
        from? Enter the name with extension (e.g., data.
        dat): ";
164     cin >> filename;
165     data histogram(filename);
166     cout << "The maximum value is: " << histogram.
        getMax() << endl;
167     cout << "The minimum value is: " << histogram.
        getMin() << endl;
168     cout << "The mean value is: " << histogram.
        getMean() << endl;
169     cout << "The standard deviation is: " <<
        histogram.getStdDev() << endl;
170     cout << "The standard error of the mean is: " <<
         histogram.getMeanDev() << endl;
171     cout << "Would you like to create a histogram? (
        Yes=Y, No=N)" << endl;
172     cin >> response;
173     if(response == 'Y' || response == 'y') {
174         cout << "Into how many intervals would you
        like to divide the data?" << endl;
175         cin >> intervals;
176         histogram.createHistogram(intervals);
177         for (int i = 1; i <= intervals; i++) {
178             cout << "Bin number " << i << " contains
        " << histogram.showHistogram(i) << " elements."
        << endl;
179         }
180     }
181     system("PAUSE");   // Command for Windows cmd.exe
        console
182     return 0;
183 }
```

Code 3.6 Implementation of a C++ class managing the problem of a histogram computation

3.3 Python Class

In Python, everything is an instance of a class. You use the class keyword to create
a new class, followed by the class name and a colon (:). The body of the class then
defines its variables, constructor, and methods. These methods are similar to those
in C++, where they are implemented as functions.

As with functions, a class must be defined before it can be used in the code.

To make a parallel with the C++ class, we'll try to implement in Python the C++ Code 5.1:

```python
class operation:
    first =1
    second = 1
    def __init__(self,a,b):
        self.first= a
        self.second= b
    def changeFS(self,a,b):
        self.first= a
        self.second= b
    def changeF(self,a):
        self.first= a
    def changeS(self,a):
        self.second= a
    def GetF(self):
        return self.first
    def GetS(self):
        return self.second
    def Sum(self):
        return self.first +self.second
    def FminusS(self):
        return self.first -self.second
    def SminusF(self):
        return self.second -self.first
    def Product(self):
        return self.first *self.second
    def SdividebyF(self):
        return self.second /self.first
objvar = operation(2,3)
print(objvar.Sum())
print(objvar.SdividebyF())
objvar.first =20
print(objvar.GetF())
```

Code 3.7 Example of implementation of a class solving of the main mathematical operations

Proceed by commenting on the code to highlight similarities and differences.

The class is implemented from lines 1 to 23 of Code 3.7. At lines 2 and 3, the internal variables of the class are defined, followed by the implementation of all methods, structured similarly to how we learned to define functions.

The constructor, which must be named __init__, is implemented at lines 4-6. In this case, it is also used to receive the values the user wants to pass when creating an instance of the class.

Another essential aspect to highlight is using the self variable in method implementations. It refers to the instance itself, enabling access to class attributes. While "self" is the conventional name for readability, it is not a strict requirement; any other name could be used. However, adhering to this convention is strongly recommended for clarity and consistency. Lines 31 and 32 of Code 3.7 illustrate that, in Python, all variables and methods are always public. This is evident in how we first modify the variable through an instance, as demonstrated when we print its values in line 32 of the code.

3.3.1 Inheritance in Python Class

In the example shown in Code 3.8, we present a Python adaptation, with some modifications, of the C++ Code 3.4. The child classes Rectangle and Triangle are defined using the class keyword, followed by their respective names and, in parentheses, the name of the parent class Shape. Rectangle and Triangle inherit the constructor and methods of the Shape class while defining their local method area. Although this method shares the same name in both classes, since it logically performs the same operation, it is implemented using different equations specific to each geometrical shape.

```python
class shape:
    base =1
    height = 1
    def __init__(self,a,b):
        self.base= a
        self.height= b
    def changeBH(self,a,b):
        self.base= a
        self.height= b
    def changeB(self,a):
        self.base= a
    def changeH(self,a):
        self.height= a
    def GetB(self):
        return self.base
    def GetH(self):
        return self.height
class rectangle(shape):
    def area(self):
        return self.base *self.height
class triange(shape):
    def area(self):
        return self.base *self.height/2

objvar = rectangle(2,3)
print("showing the use of the mather class GetB",
    objvar.GetB())
print(objvar.area())
objvar = triange(2,3)
print("showing the use of the mather class GetH",
    objvar.GetH())
print(objvar.area())
```

Code 3.8 Example of inheritance in Python Class

3.3.2 Examples

Below, we present the Python implementation of a class designed to solve the physical problem of uniform circular motion. The objective of the exercise is to develop a solution that, given the radius and angular speed—both provided by user input—allows for the calculation of key motion physical quantities. The class includes methods to compute the period, tangential velocity, and centripetal acceleration, providing a structured approach to analyzing uniform circular motion.

In Code 3.9, the math Python package has access to the mathematical operations for the calculation requested in the problem.

```python
# -*- coding: utf-8 -*-
import math

class UniformCircularMotion:
    def __init__(self, radius, angular_velocity):
        self.set_radius(radius)
        self.set_angular_velocity(angular_velocity)

    # Input methods with validation
    def set_radius(self, radius):
        if not isinstance(radius, (int, float)):
            raise TypeError("The radius must be a
number.")
        if radius < 0:
            raise ValueError("The radius cannot be
negative.")
        self.__radius = float(radius)

    def set_angular_velocity(self, angular_velocity):
        if not isinstance(angular_velocity, (int,
float)):
            raise TypeError("The angular velocity
must be a number.")
        if angular_velocity < 0:
            raise ValueError("The angular velocity
cannot be negative.")
        self.__angular_velocity = float(
angular_velocity)

    # Calculation methods
    def calculate_tangential_velocity(self):
        return self.__angular_velocity * self.
__radius

    def calculate_centripetal_acceleration(self):
        return self.__radius * (self.
__angular_velocity ** 2)

    def calculate_period(self):
        if self.__angular_velocity == 0:
```

```python
33          raise ValueError("The angular velocity
    must be nonzero to calculate the period.")
34      return 2 * math.pi / self.__angular_velocity
35
36    # Getter methods for private variables
37    def get_radius(self):
38        return self.__radius
39
40    def get_angular_velocity(self):
41        return self.__angular_velocity
42
43    def __str__(self):
44        """
45        Textual representation of the
    UniformCircularMotion object.
46        """
47        return f"UniformCircularMotion(radius={self.
    __radius}, angular_velocity={self.
    __angular_velocity})"
48
49 # Example usage of the class
50 if __name__ == "__main__":
51    try:
52        # User input
53        radius = float(input("Enter the radius (in
    meters): "))
54        angular_velocity = float(input("Enter the
    angular velocity (in radians per second): "))
55
56        # Creating an object of the class
57        motion = UniformCircularMotion(radius,
    angular_velocity)
58
59        # Calculations
60        print(f"Motion object: {motion}")
61        print("Tangential velocity:", motion.
    calculate_tangential_velocity(), "m/s")
62        print("Centripetal acceleration:", motion.
    calculate_centripetal_acceleration(), "m/s^2")
63        print("Period:", motion.calculate_period(),
    "s")
64
65        # Example using setters
66        motion.set_radius(10)
67        motion.set_angular_velocity(3)
68        print(f"Motion object after setters (after
    modification): {motion}")
69        print("Tangential velocity (after
    modification):", motion.
    calculate_tangential_velocity(), "m/s")
70        print("Centripetal acceleration (after
    modification):", motion.
    calculate_centripetal_acceleration(), "m/s^2")
```

```
71        print("Period (after modification):", motion
    .calculate_period(), "s")
72    except ValueError as ve:
73        print(f"Value error: {ve}")
74    except TypeError as te:
75        print(f"Type error: {te}")
76    except Exception as e:
77        print(f"An error occurred: {e}")
```

Code 3.9 Implementation of a Python class for modeling and managing problems involving uniform circular motion

3.4 OOP in Data Analysis

When working on a large-scale software project, such as developing interfaces for data acquisition, creating programs for data monitoring and analysis, running simulations, and comparing results, object—oriented Programming (OOP) is extremely valuable. By breaking down tasks into smaller, manageable components, different project parts can be distributed among collaborators. This modular approach also simplifies maintenance, as fixing issues in a specific class does not impact the overall functionality of the project.

In most cases, building a project from scratch is unnecessary. Extensive libraries, such as those written in C++ for data analysis frameworks or Python packages, provide pre-existing solutions that can be readily assembled, like LEGO bricks, to construct the desired functionality.

A fundamental best practice in software development is to avoid reinventing the wheel. Before writing a new class, checking whether an existing one already addresses the problem is essential. If a suitable implementation is available, it should be leveraged instead of developing redundant code.

In data analysis, typical tasks include reading data stored in various formats (ASCII, binary, etc.), performing mathematical computations, and presenting results through graphs, tables, and reports. Often, rather than developing a complex program, a short script is sufficient to accomplish a specific task efficiently—whether as a throwaway code or as a partially reusable component for future applications.

Let's consider a practical example: a weather monitoring station equipped with sensors for temperature, pressure, humidity, wind speed, and light intensity. The system records measurements every ten seconds for an entire week, storing the data in an ASCIIfile where each row corresponds to a single measurement, and individual values within a row are separated by tab spaces. Suppose we want to perform a statistical analysis of the pressure measurements stored in the second column of the data matrix. Specifically, we need to extract the pressure values, compute a histogram, and visualize the results in a graph.

To achieve this, we must be able to read only the data from the second column, process it to generate the histogram, and then create a graphical representation.

The next chapter will explore how to leverage pre-existing classes efficiently. These classes will enable us, with just a single command, to load the entire dataset into a 2D array; other methods allow us to slice the desired data, others to iterate over it, to compute the histogram, and then generate a plot that can be saved in any available graphical format.

3.5 Exercises

3.1 Write a class for handling currency conversion between the euro, dollar, and Swiss franc. The class must contain methods for loading data in any currency, returning data in any currency, and storing the current exchange rates for conversions. The class must be tested in a working program with an example that proves its functionality.

3.2 Write a class that requires two real numbers as input and has implemented methods to acquire the numbers and to provide the sum, subtraction, product, and the logarithm of the first number multiplied by the second number.

3.3 Write a class that fully solves the problem of the simple pendulum, with a private variable for the length of the pendulum, and methods that return the frequency and period of the pendulum. Implement at least two constructors. Test the functionality of the class in the main program.

3.4 Implement a class that fully solves the triangles problem and can provide the area and perimeter for any triangle. Implement at least two constructors. Test the class's functionality in the main program.

3.5 Implement a class that fully solves the problem of projectile motion, accepting initial velocity and launch angle as input, and can return the range, maximum height, and the distance at which the maximum height is reached. Implement at least two constructors. Test the functionality of the class in the main program.

3.6 Write a class with a vector as a private variable and provide public methods capable of loading data, returning the number of loaded data, and determining the maximum and minimum loaded values. Sort the vector and print it sorted.

3.7 Write a class capable of fully solving the problem of rectilinear motion. It should take the starting position, velocity, and acceleration as private variables, and provide methods that return the position and velocity at a given time t greater than zero. Implement at least two constructors. Test the functionality of the class in the main program.

3.8 Write a class capable of fully solving the problem of free fall without friction. It should take the initial height as a private variable and provide methods that return the potential and kinetic energy at a given time $t > 0$ or at a certain distance travelled. Implement at least two constructors. Test the functionality of the class in the main program.

Part II
Insights and Applications

In Part II, insights and applications will be shown on what was seen previously—in Part I of this book—on basic knowledge.

Concerning frameworks and distributions in Python, we will report information and examples on Python's libraries like NumPy and pandas—used for data manipulation—and Matplotlib—that is popular for data visualization. Furthermore, a C++ framework, ROOT, developed by CERN, will be described. ROOT is used for data analysis in high-energy physics, providing tools for data processing, statistical analysis, and visualization.

Data Processing and Visualization Methods are really important for experimental physicists. In python, data is often processed using pandas for data frames, NumPy for numerical operations, and SciPy for scientific computing, instead visualization is achieved using Matplotlib. Regarding C++, data processing involves using efficient algorithms and data structures and visualization can be done using libraries like ROOT, which provides advanced plotting capabilities.

Finally, we will propose several type of measurements of physical quantities realized by means of microcontrollers or microprocessors, and sensors. All the data acquired in these experiments will be analyzed by means of all the tools reported in this book.

Chapter 4
Analysis Frameworks and Distributions

4.1 The Analysis Framework ROOT

ROOT is a data analysis framework conceived, developed, and distributed by CERN. It offers a vast collection of open-source C++ libraries designed for data analysis, simulation, data generation, storage and management of large datasets, support for online data acquisition systems, and much more.

After listing all these capabilities, it's worth emphasizing another key strength of ROOT: it is built on C++, a language widely taught in scientific and technical faculties worldwide. C++ is a powerful and versatile programming language, commonly used not only for system-level programming—such as operating systems—but also as the foundation for high-level languages like Python. And yes, we're talking about standard C++—no variants, no dialects, no special adaptations.

While there are other excellent open-source tools for data analysis, such as Scilab, Octave, and others, that we strongly encourage you to explore, they often rely on their scripting languages, which require additional learning. Remember: being a programmer isn't just about knowing a language; learning a new language becomes much easier once you are one. Mastering C++ and having direct access to a powerful platform like ROOT is a significant advantage that should not be overlooked.

Understanding C++ unlocks ROOT's full potential. ROOT is not just a user-friendly interface with colorful buttons that execute commands behind the scenes. Instead, it offers a comprehensive set of tools—and it's up to you to use them effectively. That means writing code, starting from simple scripts and evolving to more complex solutions as your needs grow.

ROOT supports several modes of use: an interactive command prompt (similar to Perl, Python, or Gnuplot), script interpretation and function execution via files with a .C extension, library integration within compiled C++ programs, and the option to compile macros before execution to boost performance in interactive use. This is made possible by Cling, ROOT's C++ interpreter.

© The Author(s), under exclusive license to Springer Nature Switzerland AG 2026

S. Vasi et al., *Open Source Tools for Physics Data Analysis*, Undergraduate Texts in Physics, https://doi.org/10.1007/978-3-032-00721-6_4

In the following sections, we'll briefly overview how ROOT works. These notes are based primarily on hands-on experience. They intentionally avoid technicalities since the official ROOT website offers thorough documentation, including class references, user guides, and beginner-friendly tutorials.

4.1.1   How to Install ROOT

Root can be installed on different operating systems: Linux, Windows, and macOS. We recommend working on a unix-based os and installing Root by downloading and compiling the source code. This procedure is less user-friendly but guarantees ROOT's best performance and stability. Installing Root on Linux is very simple. First, you must install the prerequisites, which are additional libraries, compilers, and compilation tools necessary for the correct functionality of Root. To do this, it is enough to look for "CERN root dependencies" on a search engine to find the following address: https://root.cern/install/dependencies/. Then it would help if you looked for your Linux distros or macOS version and followed the instructions.

After successfully installing all libraries, you can proceed with installing ROOT. The pre-compiled binary version of ROOT is available for different Linux, Mac OS, and Windows operating systems. Still, we recommend downloading the source code of the last stable release, unzipping the file, and then compiling it following the instructions of the README file available in the archive file. If the compiling procedure ends successfully, you are sure that all ROOT features work correctly.

4.1.2   Root Macro

After the installation of ROOT and the setup of all environment variables via this-root.sh script. You can start using the shell script. In any Unix terminal by executing the command root, a ROOT session will begin with an environment where all C++ and ROOT libraries are included and a C++ interpreter is available (see Fig. 4.1). You can use the ROOT terminal session by writing a C++ string of one or more instructions, and if they are correct, the interpreter will execute as in Figs. 4.2, 4.1.

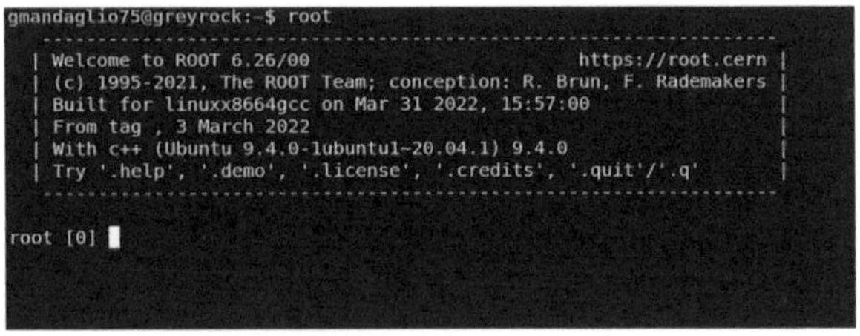

Fig. 4.1 ROOT session on the Linux terminal

Fig. 4.2 Instructions running in a ROOT session

```
gmandaglio75@greyrock:~$ root
   ------------------------------------------------------------
  | Welcome to ROOT 6.26/00                      https://root.cern |
  | (c) 1995-2021, The ROOT Team; conception: R. Brun, F. Rademakers |
  | Built for linuxx8664gcc on Mar 31 2022, 15:57:00 |
  | From tag , 3 March 2022 |
  | With c++ (Ubuntu 9.4.0-1ubuntu1~20.04.1) 9.4.0 |
  | Try '.help', '.demo', '.license', '.credits', '.quit'/'.q' |
   ------------------------------------------------------------
root [0] float a; float b;
root [1] a=10;
root [2] b=3;
root [3] cout<< a + b<<endl;
13
root [4] TVector3 vect(a,b,2.1);
root [5] cout <<"printing the vector magnitude "<<vect.Mag()<<endl;
printing the vector magnitude 10.6494
root [6] cout <<"printing the vector polar angle in radiant "<<vect.Theta()<<endl;
printing the vector polar angle in radiant 1.3723
root [7]
```

In the example reported in Fig. 4.2, two float variables are declared, then values are assigned, a TVector3 ROOT class is declared and initialized by its constructor and finally, the object of the TVector3 is used to obtain the magnitude and the polar angle in radians. The interpreter can help test pieces of code or study the functionality of some classes, but as in Python, the best way to use Root is to write a macro code and run it using its interpreter. The simplest way to create a macro is to write all the C++ instructions enclosed between two simple curly brackets in a text file with the extension dot C-capital "nameofthemacro.C". Then, to interpret the macro, it is enough in the terminal to write "root nameofthemacro.C" in the macro folder. For example, we can transport the code we tested in the previous example in a macro file called test.C in the following way:

```
1  {//start the root macro - c++ code
2  float a=10, b=3;
3  cout<<a<<"  "<<b<<endl;
4  TVector3 vect(a,b,2.1);
5  cout<<"printing the vector magnitude"<<vect.
       Magnitude()<<endl;
6  cout<<"printing the vector polar angle in radians "
       <<vect.Theta()<<endl;
7  }//end the macro
```

Code 4.1 Example of macro replacing the effect of commands digited on the interpreter console

Another way to run the macro test.C is to open a Root session and write ".x test.C" on the root interpreter.

4.1.3 Functions TF1, TF2, T3

Root provides a way to represent mathematical functions through the TF1, TF2, and TF3 classes, also in a parametric form, which is valuable for fitting data by minimizing procedures. We invite readers to use the public documentation on the Root website for details on the class method. Root is a living project, so improvements and changes can happen.

Let's imagine that we have learned about the damped oscillations phenomenon in the lesson and want to practice with the parameters that rule this phenomenon. We could use TF1 in the following way:

```cpp
{//Macro function.C
{
//not-parametric functions examples
TF1 *expy = new TF1("expy","10*exp(-x/3)",0,15);
TF1 *expyneg = new TF1("expyneg","-10*exp(-x/3)"
    ,0,15);
//parametric function example
TF1 *expsin = new TF1("expsin","[0]*exp(-x/[1])*sin
    ([2]*x)",0,15);
//method to change the values of the parameters
expsin->SetParameters(10,3,4);
expy->SetLineColor(1);
expy->SetLineWidth(2);
expyneg->SetLineColor(1);
expyneg->SetLineWidth(2);
expsin->SetLineColor(4);

expsin->Draw();
expy->Draw("same");
expyneg->Draw("same");
}//Macro end
```

Code 4.2 Root Macro drawing the functions describing the damped oscillations phenomenon

In this simple script, we see how to write a complete function inside the constructor in the object *expy* and how to write it depending on parameters ([0], [1], etc.) and how to pass the values to the *expsin* function through the *SetParameters* method. The execution of the macro produces the following Fig. 4.3.

Let's try to write a slightly more complicated function. Imagine a phenomenon that obeys a law of the form:

Fig. 4.3 Result of the macro *function.C* execution

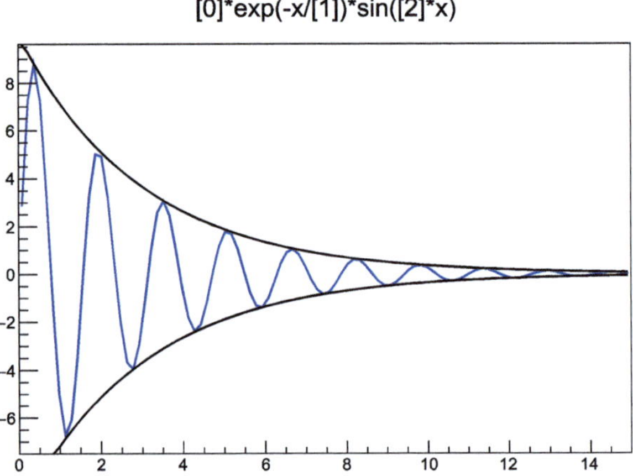

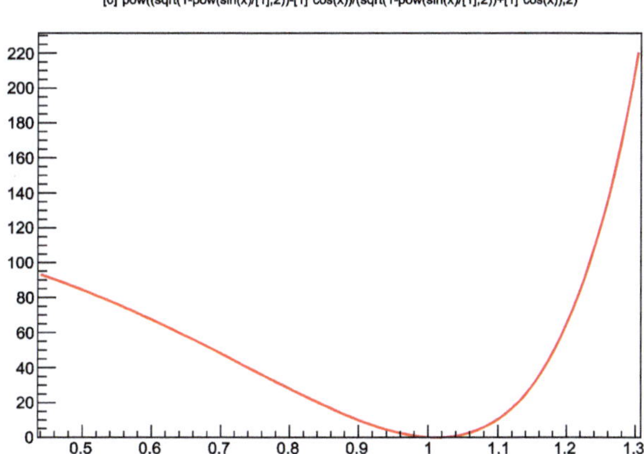

Fig. 4.4 Graphical representation of the Complicatedfunction.C macro

$$R_p = A \cdot \frac{\sqrt{1 - (\frac{sin(\theta)}{n_1})^2} - n_1 \cdot cos(\theta)}{\sqrt{1 - (\frac{sin(\theta)}{n_1})^2} + n_1 \cdot cos(\theta)} \tag{4.1}$$

Let's write the script capable of plotting the function:

```
{//Macro Complicatedfunction.C
{   //script funzione_complicata.C
  TF1 *ff=new TF1("ff","[0]*pow((sqrt(1-pow(sin(x)
    /[1],2))
                         -[1]*cos(x))/   (sqrt(1-pow(
    sin(x)/[1],2))
                         +[1]*cos(x)),2)"
    ,25*3.14/180,75*3.14/180);
  ff->SetParameter(0,2320);
  ff->SetParameter(1,1.6);
  ff->Draw();
}//Macro end
```

Code 4.3 Root defying a parametric function describing phenomenon and drawing it

The result of the execution of the macro produces the following Fig. 4.4.

We described the second model as more complex than the previous one due to the mathematical function it involves. For instance, suppose we assign the value $n_2 = 0.5$ in Eq. 4.1. What would happen? From a mathematical perspective, the sine function is bounded within the interval ± 1. Therefore, if $n_2 < 1$, there may be specific values of θ for which the function does not yield real results. To explore this scenario, let us run the script by assigning the value 0.5 to the parameter using the command `ff -> SetParameter(1, 0.5);`. The resulting output will be as follows:

```
Processing Complicatedfunction.C...
Info in <TCanvas::MakeDefCanvas>:  created default TCanvas with name c1
Warning in <TCanvas::ResizePad>: Inf/NaN propagated to the pad.
Check drawn objects.
Warning in <TCanvas::ResizePad>: c1 height changed from 0 to 10

root [1] Warning in <TCanvas::ResizePad>: Inf/NaN propagated to the pad.
Check drawn objects.
Warning in <TCanvas::ResizePad>: c1 height changed from 0 to 10

Warning in <TCanvas::ResizePad>: Inf/NaN propagated to the pad.
Check drawn objects.
Warning in <TCanvas::ResizePad>: c1 height changed from 0 to 10
```

ROOT produces error messages and prints them on the screen, but we made the error. This does not mean that ROOT is not working; on the contrary, it means the opposite. This is to remember that while software tools help us achieve results, we remain responsible for them.

4.1.4   Fitting Procedure

Fitting is a mathematical and numerical procedure used to determine the parameters of a chosen function—often referred to as a model or law—that best describes the behavior of a set of correlated data and the associated uncertainties in those parameters. For instance, if we measure the elongation of a metal wire under increasing tension (produced by progressively heavier weights), we might assume a linear relationship between stress and elongation. Similarly, when heating a metal bar, we can relate temperature increase to thermal expansion, or in the case of a swinging pendulum, the oscillatory motion can be modeled using a sinusoidal function. These are examples where fitting is applied to physical laws. However, the underlying model doesn't have to be physical—it can also be statistical. For example, the lengths of nails manufactured for carpentry, or the plastic balls mentioned in earlier sections, tend to follow a Gaussian distribution around a mean value. In such cases, the histogram of the measured values can be fitted with a Gaussian function. Thus, fitting is a versatile tool that applies to physical phenomena and statistical trends.

Root provides a series of built-in functions widely used for performing fits, or it allows the user to write their law for fitting using a class of type **TF1** and use it in a completely customized calculation.

The built-in functions in root include the Gaussian (keyword: **gaus**, Gaussian with 3 parameters: $f(x) = p0 \cdot exp(-0.5 \cdot ((x - p1)/p2)^2)$), the Landau (**landau**, a function with two parameters, mean and sigma), the exponential function (**expo**, exponential function with two parameters: $f(x) = exp(p0 + p1 \cdot x)$ f(x) = exp(p0+p1·x)), and polynomials (**polN**, where N should be replaced with 0 for fitting a constant, 1 for a straight line, 2 for a parabola, and so on—where N indicates the polynomial order desired).

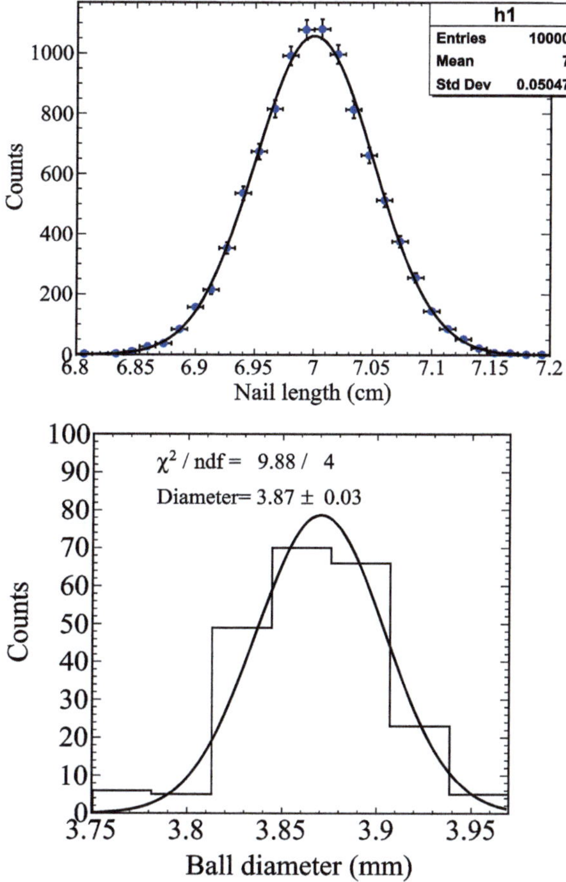

Fig. 4.5 Data analysis examples, measurements of length and the diameter of nails (top panel) and balls (bottom panel)


```
//data is an object of the class TH1F
data->Fit("gaus");
```

and we would obtain the result shown in Fig. 4.5. This is merely an introductory example to familiarize you with these methods. The Fit method has many options that will allow you to study your data in a much more sophisticated way (https://root. cern.ch/root/html534/guides/users-guide/FittingHistograms.html#the-fit-method).

The distribution is precise in the examples just presented, and by directly invoking the fitting method, we have a good chance of obtaining the desired result. In general, when performing fitting operations on a computer, the parameter search algorithms need a suggestion—they need us to provide values close to the expected ones. Then we delegate the task to them to find the best values. 'Best' in the sense of those that produce the curve that most closely matches the data. The values provided as starting points for the fit are called fit initialization parameters. One method for assessing a fit's quality is the χ^2 method. The χ^2 variable is constructed as the sum of the squared deviations between the experimental values and those of the fit, normalized by the square of the sum of the respective errors (experimental error and uncertainty of the fitting function). Finding the function's parameter values that minimize this variable can be a criterion for finding the best fit.

4.1.5 TRandom Classes and Toy MonteCarlo

Generating synthetic data that follows a specific probability distribution is often essential in data analysis. This is important for testing algorithms, assessing the feasibility of experiments, aiding data interpretation, or validating statistical models. Complex simulation frameworks typically begin with code that generates events, and every event generator depends on a pseudo-random number generator.

ROOT provides dedicated classes, such as the TRandom* series, designed to generate pseudo-random numbers with varying balances of quality and speed. These include TRandom (now deprecated), TRandom1 based on the RANLUX algorithm, TRandom2 which uses L'Ecuyer's Tausworthe generator, and TRandom3, which implements the Mersenne Twister generator—known for its speed and very long period—among others.

The use of these tools will be explored in greater detail in the next chapter.

4.2 Python Distributions

Python distributions are integral parts of the Python ecosystem. They allow you to work quickly and effectively in data science by containing everything you need to start writing code.

A Python distribution is a versioned and compressed archive that contains Python libraries (i.e., application packages) that you download using a package manager like pip. Some popular Python distributions include:

- *Anaconda*, a free and open-source distribution of Python and R for scientific computing.
- *ActivePython*, a commercially supported, pre-compiled distribution of Python.
- *Python (x,y)*, a scientific-oriented Python Distribution based on Qt, Eclipse, and Spyder.

Usually, Python distributions are also platforms providing an "environment" for developers to write, test, and deploy their Python applications. They often come with tools for package management, environment management, and more. Examples of platforms are again Anaconda and *Jupyter Notebook*.

Depending on their needs, users might choose different distributions and platforms to build a Python project. For instance, if you're new to Python, you might start with a pre-built version such as Anaconda. If you're an experienced "Pythoner", you might prefer to install all the packages you require using pip manually.

In this section, we will introduce Jupyter Notebook and JupyterLab, Anaconda, and show their support for writing and testing code in data analysis and scientific computing. In the following, it is possible to find more information about:

- Anaconda.
- Jupyter Notebook.

4.2.1 Jupyter

Jupyter is a project that supports over 40 programming languages, including Python, and provides free software, open standards, and web services for interactive computing and data science. In particular, JupyterLab and Jupyter Notebook are web-based, interactive development environments that represent the ultimate expression of this project's ideas.

4.2.1.1 Jupyter Notebook

Jupyter Notebook is a web-based interactive computing platform that allows you to create and share documents that contain live code, equations, visualizations, and narrative text. In fact, one of the biggest advantages of using notebooks is that they allow you to combine documentation, code, and results into one interactive environment. This makes it ideal for data exploration and presentation of scientific results, generating a rich, interactive output that can be composed of elements such as HTML, images, videos, LaTeX, and so on. Furthermore, notebooks can be shared with others using email, Dropbox, GitHub, etc.

Installation and first start

The installation of Jupyter Notebook can be easily managed via pip—using the command `pip install—user notebook`—in any operating system. After that process is finished, it is possible to start the notebook by entering on the command line the string `jupyter-notebook`.[1]

---

[1] Of note is that there is a possibility that the notebook may not start, indicating "command not found". In this case, adding the Jupyter notebook directory to the PATHs allowed by the operating system is necessary. This procedure is beyond the scope of this book, and we recommend that you investigate this further on the Web or, for example, via the following URL: https://python.land/data-science/jupyter-notebook.

Fig. 4.6 First interface of Jupyter Notebook

First use

If you start Jupyter Notebook, a web page in the browser opens automatically (http://localhost:8888/). As shown in Fig. 4.6, it is possible to create a new Notebook by clicking on the "New" button and then on "Python 3" (if you have installed this version) or File → New → Notebook. In the new window, we will finally find the interface of a Jupyter notebook. It consists of a menu bar and a toolbar. Code can be organized in numbered "cells" that can be executed independently by pressing "Shift+Enter" on the keyboard or clicking the $\triangleright$ "run" button in the ribbon at the top of the notebook. If you want to run all cells in a notebook automatically, go to the "Run" tab in the menu bar and select "Run All Cells" (or the relevant option) or click the $\triangleright\triangleright$ "Restart the kernel and run all cells" button in the toolbar of the notebook. When you run a cell, its content is executed, and any output appears below it. Cells are identified as either (i) Markdown, exploited to write text and/or typesetting mathematical equations, (ii) Code, used to edit and write new code, and perform tasks like loading data, plotting data and running analyses, or (iii) Raw, employed to render different code formats into HTML or LaTeX. Here, we will focus on Code and Markdown Cells.

Markdown cells

The procedure for configuring a cell as a markdown one is:

- Insert a new cell and select it or a new one;
- Use the drop-down menù on the toolbar, selecting "Markdown" or click 'm' on the keyboard.

To edit the cell, double-click inside it. The cell format has changed to allow for editing. Below are reported useful tools for formatting markdown cells.

1. Title/Headings
   To create notebook titles and section headings, use the symbol # followed by a blank space:

- # for titles;
- ## for major headings;
- ### and #### for sub-headings.

2. Emphasis
   To apply emphasis to your text:

   - Bold text: `__string__` or `**string**`;
   - Italic text: `_string_` or `*string*`.

3. Mathematical symbols
   Enclose mathematical symbols by using the symbol $, e.g., `$ mathematical symbols $`.

4. Monospace font
   Confine text using a grave accent '(a back single quotation mark). Note that the monospace font is suitable for displaying file paths, file names, message text visible to users, or text entered by users.

5. Line breaks
   Force a line break by using `<br>`.

6. Bullets
   Make use of the following methods to create a circular bulleted list.

   - A hyphen (-) followed by one or two spaces, for example: - Bulleted item
   - A space, a hyphen (-) and a space, for example: - Bulleted item
   - An asterisk ∗ followed by one or two spaces, for example: ∗ Bulleted item

   To create a sub-bullet, press `Tab` before entering the bullet point using one of the methods described above.

Important:
Ensure that each bullet point appears on a separate line.

7. Numbered lists
   Insert `1.` followed by a space to create a numbered list, for example:

   - `1.` Numbered item -> 1. Numbered item
   - `1.` Numbered item -> 2. Numbered item

   Note that the use of `1.` before the text is done, just to make things simple. The list will be numbered correctly when you run the cell. To create a substep, press `Tab` before entering the numbered item.

8. Horizontal lines
   A horizontal line will be generated if three asterisks ∗ ∗ ∗ are inserted in a new line.

```
        We propose a simple code to calculate the modulus of force using Newton's second law (knowing mass and acceleration imprinted
        on it):

In [ ]:  #Define the values of mass and acceleration
         m = 5 #kg
         a = 2 #m/s^2

In [ ]:  # The second principle of dynamics states that force (F) equals mass (m) times acceleration (a). The modulus will be:
         F=m*a #N

In [ ]:  F

In [ ]:  print(r"The value of the force having m =", m, "kg and a =", a, "m/s\u00b2 is", F, "N")
```

Fig. 4.7 Code cells in Jupyter Notebook. This simple example shows the code used for calculating Newton's Second Principle of Dynamics

9. Graphics

   In Markdown cells, you can attach image files (e.g., graphs, etc.) by dragging and dropping them into the cell.
10. External links

   Enclose an URL in two underscores _ on each side to link to an external site:
   ___[link text](URL)___link text

To see the formatted version, run this cell. Markdown cells are used for text and help describe code operations in subsequent cells. For a comprehensive list of text formatting options in Markdown, go to the "Help" tab in the JupyterLab menu bar and select "Markdown Reference".

   References:

– Markdown Cells—Jupyter Notebook 7.2.1 documentation—Read the Docs.
– The Complete Guide to Markdown Tables for Organized Data—Link.
– Master Markdown Tables: A Complete Guide & Tips—Link.
– Format Text In Jupyter Notebook With Markdown—Link.
– IBM markdown jupyter cheatsheet—Link.

Code cells

Code cells are identified by an In [ ] located at the left of each cell. Once a new line is created in the notebook, it is automatically created as a code cell. There is always the option of changing the cell type by selecting an old cell and then "Code" from the drop-down menu in the toolbar. In the following figures, we present an example of comparing a notebook before and after its compilation/execution. Figure 4.7 shows a simple example of calculating Newton's Second Principle of Dynamics[2] written in Python, illustrating us that the notebook has not been run yet. In fact, the [ ] symbol to the left of each code cell describes the state of the cell, which means that the cell has not been run yet. Furthermore, note that the first line of the code reported

---

[2] Newton's second law of motion can be formally stated as follows: The acceleration of an object as produced by a net force is directly proportional to the magnitude of the net force, in the same direction as the net force, and inversely proportional to the mass of the object.

We propose a simple code to calculate the modulus of force using Newton's second law (knowing mass and acceleration imprinted on it):

```
In [1]:  #Define the values of mass and acceleration
         m = 5 #kg
         a = 2 #m/s^2

In [2]:  # The second principle of dynamics states that force (F) equals mass (m) times acceleration (a). The modulus will be:
         F=m*a #N

In [3]:  F
Out[3]:  10

In [4]:  print(r"The value of the force having m =", m, "kg and a =", a, "m/s\u00b2 is", F, "N")
         The value of the force having m = 5 kg and a = 2 m/s² is 10 N
```

Fig. 4.8 Output of the code cells' example reported in Fig. 4.7

in Fig. 4.7 is configured as a "Markdown" cell (see paragraph "Markdown cells"). Selecting "Run All Cells", the source code is compiled and the result of this process is illustrated in Fig. 4.8.

It is highlighted that a change has also occurred to the text In [ ] . In fact, during or after the start of the code compilation, the following may appear inside the square brackets:

*: means that the cell is currently running;
1: means that the cell has finished running and was the first cell run (the number indicates the order that the cells were run in).

This example also allows us to introduce different ways of displaying outputs in Jupyter Notebook. As shown in Fig. 4.8, the lines of code In [3] and In [4] produce two different types of output. The first one, obtainable by just writing the variable name, provides a line of output identified by Out[3] with the value of the variable (in our example, the numeric value equal to 10). In [4] , instead, contains the standard Python Iprint() function, but this code doesn't put anything on the Out[ ] list in Jupyter Notebook.

Another essential feature is the ability to access the previous cells using the In and Out arrays. In the example above, Out[3] gives us the output of the expression in the In [3] cell. We can easily call this output in our Jupyter Notebook by writing Out[3] in another code cell. For example, consider that you want to use a mass that is no longer 5 kg but four times larger (i.e., 20 kg). Instead of changing the value of the variable "m", simply multiply the output 3 by 4 by writing the text Out[3]*4 in the new line of code.

4.2.1.2   JupyterLab

JupyterLab is a next-generation web-based user interface for Project Jupyter, consisting of an interactive development environment for notebooks, code, and data.

It allows working with documents and activities such as Jupyter notebooks, text editors, terminals, and custom components in a flexible, integrated, and extensible/customizable interface. Also including Jupyter Notebook, the code can produce rich, interactive output, such as, HTML, images, videos, LaTeX, and custom MIME types, supporting over 40 programming languages and understands many file formats (images, CSV, JSON, Markdown, PDF, Vega, Vega-Lite, etc.). Furthermore, you can leverage big data tools like Apache Spark from Python, R, etc.

Installation and first start

JupyterLab can be installed on your personal computer using the command `pip install --user jupyterlab` . After the installation, we can start the environment by entering on the command line the string `jupyter lab` .

Figure 4.9 shows the JupyterLab interface. It comprises the following main elements:

- Work Area.
  The main work area features a new tab-based layout that allows you to combine different documents (notebooks, data files, HTML, PDFs, images, etc.) and coding tools (terminal, code consoles, text editor, etc.) into panels of tabs that can be resized or subdivided through drag and drop. This tab-based system, absent in

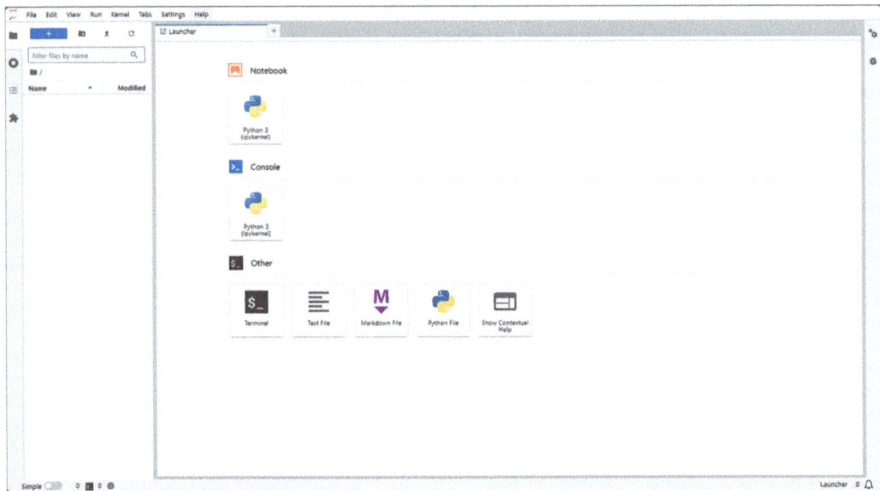

Fig. 4.9 The interface of JupyterLab

Jupyter Notebook, enables more integrated and efficient programming, allowing developers to perform most of their tasks without leaving JupyterLab.

- Left Sidebar.
  The collapsible sidebar contains various commonly-used tabs, including:

  - *File Browser*, which manages files and folders on your computer.
  - *Tab and Kernel Manager*, providing a list of tabs in the main work area and of running kernels and terminals.
  - *Table of Contents*, an interactive index based on headings included in markdown cells.
  - *Extension Manager*, allowing you to install new applications and plugins to enhance your JupyterLab experience.

- Right Sidebar.
  A second collapsible sidebar that contains:

  - *Property Inspector* that permits inspection of the information and properties of the cells closely.
  - *Debugger* which allows you to debug notebooks, code consoles, and files directly within JupyterLab.

- Menu Bar.
  The menu bar at the top of JupyterLab includes top-level menus with actions available in JupyterLab. The different menus are:

  - File: Actions related to files and directories.
  - Edit: Actions related to editing documents and other activities.
  - View: Actions that change the appearance of JupyterLab.
  - Run: Actions for running code in different activities.
  - Kernel: Actions for managing kernels.
  - Tabs: A list of the open documents and activities in the work area.
  - Settings: Common settings and an advanced settings editor. One example is the ability to choose different themes for the interface, including Dark Mode.
  - Help: A list of JupyterLab and kernel help links.

First use

Inside the "Launcher" tab (see Fig. 4.9), we can select—and, consequently, start— the project we like, such as a Jupyter notebook. Of course, we also have the option of opening an existing notebook and, by doing so, opening a new tab in the Work Area. Considering the Python 3 console or the Notebook as possible choices, we may follow all the indications and guides given previously in the book to write and compile the code also inside JupyterLab.

4.2.2 Anaconda

Anaconda is a free, cross-platform, and open-source distribution of the Python and R programming languages that aims to provide everything you need to get started with Python (or R) in one package, and to simplify package management and deployment. In fact, Anaconda comes with "Anaconda Navigator", a desktop application included in the distribution, lets you manage your applications, packages, and environments without using the command line, and a package manager called "conda" allowing to install specific versions of Python and R, and also other packages by using the command line. Furthermore, Anaconda provides an easy way to create and manage separate *environments*, self-contained, isolated[3] spaces where you can contain files and install specific versions of software packages, including dependencies, libraries, and Python versions.

4.2.2.1   Anaconda Installation

The steps to follow to install Anaconda are:

1. Download the installer.
   Go to the Anaconda Distribution page and download the installer for your operating system (Windows, macOS, or Linux).
2. Run the Installer:

   - Windows: Double-click the downloaded ".exe" file and follow the instructions on the screen. If you're unsure about any settings, you can accept the defaults.
   - macOS: Open the downloaded ".pkg" file and follow the on-screen instructions.
   - Linux: Open a terminal and run the following command to install Anaconda:

```
bash ~/Downloads/Anaconda3-<INSTALLER_VERSION>-
    Linux-x86_64.sh
```

Code 4.4 Anaconda installation on a Linux operating system

   Note that you have to replace <INSTALLER_VERSION> with the file name you downloaded.

3. Follow the Installation Prompts.
   During the installation, you can choose whether to add Anaconda to your PATH environment variable. It is generally recommended to do so.

---

[3] Isolation helps avoid conflicts between package versions and ensures that your projects have the exact libraries and tools they need. This can be very useful when working on projects with different requirements.

4. Verify the Installation:
   Open a terminal or Anaconda Prompt and type:

```
conda --version
```

Code 4.5 Verify the Anaconda installation

This should display the version of Conda installed, confirming that Anaconda is installed correctly.

5. Launch Anaconda Navigator:
   You can now launch Anaconda Navigator from your Start menu (Windows) or Applications folder (macOS). On Linux, you can start it by typing "anaconda-navigator" in the terminal.

4.2.2.2 Anaconda First Use

After the installation, you can launch "Anaconda Navigator". As already stated, it is a graphical user interface that allows you to manage packages and environments and launch applications like Jupyter Notebook.

Initially, we may want to create a new environment. To do this, go to the "Environments" tab in Anaconda Navigator, and create a new environment. This procedure facilitates the management of dependencies for different projects and permits you to choose the Python version and the packages you need.

We can use the "Environments" tab to search for and install packages in this frame. Note that there is a chance to install a package using the command line with `conda install package-name`.

From the "Home" tab in Anaconda Navigator, applications like Jupyter Notebook can be launched.

So, following the information and the example reported in Sect. 4.2.1.1, we can write a simple Python program in Jupyter Notebook via Anaconda:

```python
print("Hello, Anaconda!")
```

Code 4.6 A simple example of writing a Python code in Anaconda

4.3 PyROOT

According to the description on the root manual website (https://root.cern/manual/python/), PyROOT allows the creation of bindings between Python and C++ dynamically and automatically. It allows the use of ROOT's powerful libraries combined with Python data-science libraries like numpy, pandas, and numba.

Installing PyROOT is quite straightforward. If you install it by compiling the source code, once the prerequisites are in place, you can use ROOT directly within a Python session. Since Python is one of ROOT's prerequisites, a full build will automatically link ROOT with the Python version installed on your system.

After running the script thisroot.sh (or thisroot.fish,csh for other shells) on a Unix-like operating system—such as a Linux distribution or Apple's macOS—or the thisroot.bat batch file on Windows, ROOT will be properly configured in your environment.

We recommend installing all the required dependencies and then compiling ROOT from source. However, it is also possible to install ROOT using a package manager. On Linux, for example, you can use Snap (sudo snap install root-framework), while on macOS you can use Homebrew (brew install root) or MacPorts (sudo port install root6).

Note, though, that with package manager installations, you typically won't be able to load ROOT into a standard Python 3 session with a simple import ROOT. Instead, you'll need to launch ROOT's own Python environment using the PyROOT command from the terminal, which starts a Python interpreter bundled with ROOT.

We strongly recommend building ROOT from the source for full functionality and compatibility. This ensures that, if the compilation succeeds, all the library's features will work correctly.

Firstly, the root is importable directly from Python; in the second way, you must use the interface PyROOT.

4.4 Exercise

4.1 Install each data analysis framework described in the chapter and test its functionalities.

Chapter 5
Data Elaboration and Visualization

5.1 Intro

As already stated, effectively manipulating and visualizing data is essential in the modern era of scientific research and data-driven decision-making. Data manipulation allows researchers and analysts to clean, transform, and prepare raw data for analysis, ensuring accuracy and relevance. After that, visualization plays a crucial role in interpreting complex datasets, revealing patterns, trends, and insights that might otherwise remain hidden.

This chapter introduces key open tools used for these tasks: ROOT, a robust framework developed by CERN; Gnuplot, a flexible plotting utility; and Python libraries such as Matplotlib and Pandas, which offer versatile solutions for handling and visualizing data.

5.2 Data Elaboration and Visualization with Root

5.2.1 Where Lives the Figures: TCanvas

The first class that must be used to manage a graphical representation of data analysis in the ROOT framework is TCanvas. This class provides many features to control the quality of the figure. In the example reported in Fig. 5.1 we represented by arrows and labels how it is possible to control the highness (W_H) and the witness (W_H) of the canvas, the distances between the figure axes and the boundaries of the canvas, the left (W_L), the right (W_R), the bottom (W_B) and the top (W_T). The height and width of the canvas can be defined directly by using the constructor of the class, and the graphical units are the point pt, while for the other distances, dedicated methods of the class must be used, as reported in the following code:

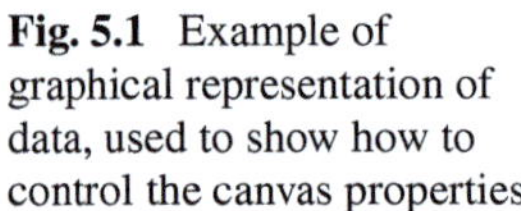

Fig. 5.1 Example of graphical representation of data, used to show how to control the canvas properties

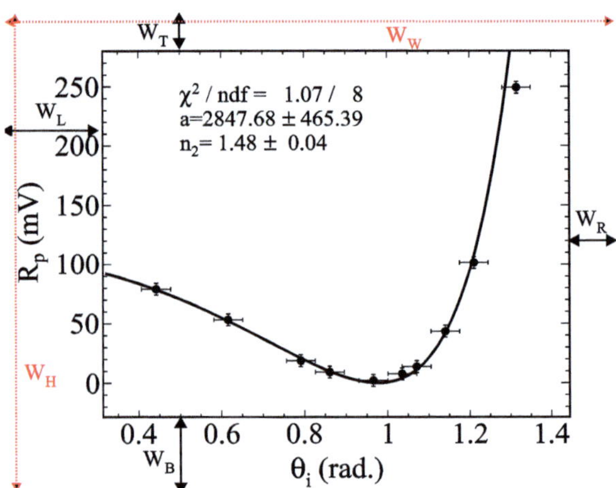


```
 1 {//start the root macro - c++ code
 2 TCanvas *Box = new TCanvas("Box", "Box",600,500);
 3 //the witness is 600 pt while the highness is 500
 4 Box->textgreaterSetFillColor(0); //0 means color background is
       white
 5 Box->textgreaterSetBorderMode(0);
 6 Box->textgreaterSetBorderSize(2);
 7 Box->textgreaterSetLeftMargin(0.16);//left margin is 16% of
       the witness of the figure
 8 Box->textgreaterSetRightMargin(0.08);//right margin is 8% of
       the figure
 9 Box->textgreaterSetTopMargin(0.08);//top margin is 8% of the
       highness
10 Box->textgreaterSetBottomMargin(0.16);//bottom margin is the
       16%
11 }//end the macro
```

Code 5.1 The use of TCanvas class

The ROOT user guide exhaustively explains the features with convenient graphical interfaces, but the goal is only to introduce the reader to the huge possibilities the analysis framework offers.

5.2.2 An Easy Way to Represent Data: TGraph and Relatives Classes

Visualizing data through graphs is one of the most common tasks in data analysis. Often, this is done using proprietary commercial software with user-friendly graphical interfaces, as it is perceived to be the most straightforward approach. While this simplicity is undeniable, it comes with financial costs and inherent lim-

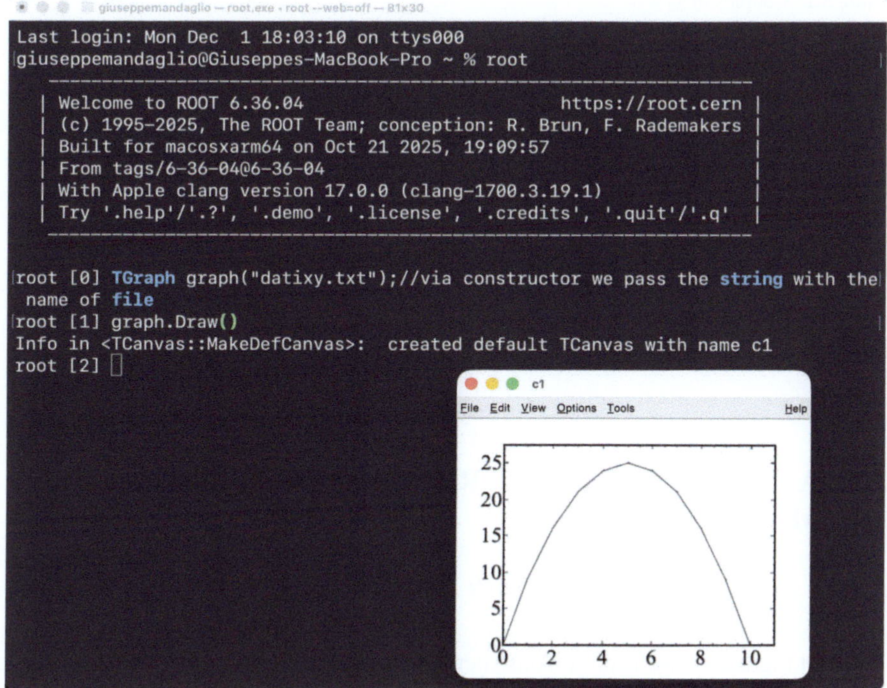

Fig. 5.2 Representation of data with TGraph class using the Root interpreter

itations. The data analysis framework ROOT offers the TGraph class, along with
its extensions TGraphErrors and TGraphAsymErrors, which enable straightforward
data graphing—provided you have some basic knowledge of C++.

As the first example, let's try to represent in a graph the couple of x, y-data points
reported in Table 5.1: We can use the interpreter directly as in Fig. 5.2: or we should
write the code in a macro as in the following Code 5.2.

```
{//start the root macro - c++ code
TGraph graph("dataxy.txt");
graph.Draw();
}//end the macro
//to run it is enough to write on the terminal root namemacro.
  C
```

Code 5.2 Root macro to plot a table of x, y-data

Table 5.1 Couples of x, y data points saved in a text file dataxy.txt

0	0
1	9
2	16
3	21
4	24
5	25
6	24
7	21
8	16
9	9
10	0

5.3 A Convenient Statistical Representation of Repeated Measures: TH1 Classes

5.3.1 *My First Histogram with Root*

The first ingredient we need to construct a histogram is the data. Let's assume these have been copied into a file called dati1x.dat and written in a single column, or a single row, or disorderly; in our case, it doesn't matter.

The simplest way to compute a histogram with ROOT is to write a macro capable of reading the data, providing it to an object of the TH1D class, and then using the Draw method to visualize the result. The Code 5.5 that accomplishes this is shown below:

```
1  {//Macro "MyFirstHist.C"
2  ifstream reader("data.dat");
3  if(reader.fail())
4      {
5      cout<<"the file does not exist!"<<endl;
6      }
7  double a;
8  //TH1D object declaration
9  TH1D *hist = new TH1D("hist","",50,0,0);
10 while(reader>>a)
11     {
12     hist->Fill(a);
13     }
14 hist->Draw();
15 cout<<"the mean value of data is "<<hist->GetMean()<<endl;
16 cout<<"the standard deviation of data is"<<hist->GetRMS()<<
       endl;
17 }//Macro end
```

Code 5.3 Root Macro computing an histogram

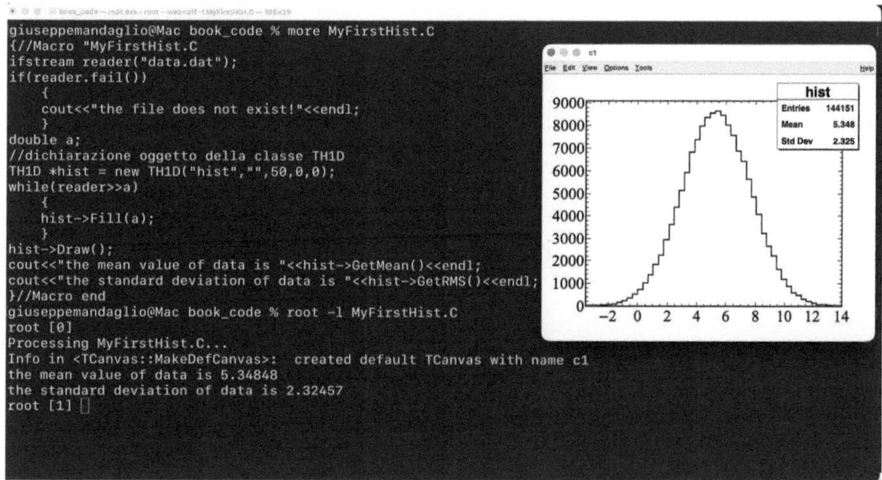

Fig. 5.3 Execution of "MyFirstHist.C" macro by Root

How the macro works is shown in Fig. 5.3.

As illustrated in Fig. 5.3, ROOT has launched a graphical interface that displays the histogram image. By default, the first canvas automatically created by ROOT is named c1. Later on, we will explore managing canvases entirely using the TCanvas class . Suppose you wish to save the generated figure for use in an article, report, or presentation. In that case, you can use the canvas's SaveAs() method, as demonstrated in the code snippet in Fig. 5.4.

```
//Code fragment to be added to "MyFirstHist.C" macro
hist->Draw();
c1->SaveAs("Myhist.pdf");//the extension of the name file
    tells to the TCanvas object how to save the file
}//end of the macro
```

Code 5.4 Code snippet demonstrating how ROOT saves a figure in PDF format

This serves as a clear demonstration of the power of object-oriented programming. We addressed the same problem in Sect. 2.3.6.7, but the code was significantly more complex. In this case, the main challenge lay in reading and passing the data to the class object. The hist object in the macro example generates a graphical representation of the data as bar histograms, but it also offers a wide range of features through its many methods. For instance, the *GetMean()* method returns the average value, while *GetRMS()* (Root Mean Square) provides the standard deviation of the data sample. Readers are encouraged to implement procedural code in C++ to compute the mean and standard deviation and better appreciate this programming paradigm's advantages.

```
giuseppemandaglio@Mac book_code % root MyFirstHist10.C
  ------------------------------------------------------------------------
  | Welcome to ROOT 6.36.04                          https://root.cern |
  | (c) 1995-2025, The ROOT Team; conception: R. Brun, F. Rademakers |
  | Built for macosxarm64 on Oct 21 2025, 19:09:57                   |
  | From tags/6-36-04@6-36-04                                        |
  | With Apple clang version 17.0.0 (clang-1700.3.19.1)              |
  | Try '.help'/'.?', '.demo', '.license', '.credits', '.quit'/'.q'  |
  ------------------------------------------------------------------------

root [0]
Processing MyFirstHist10.C...
the mean value of histogram 1 is 4.24196
    *****************
the mean value of histogram 2 is 0.497531
    *****************
the mean value of histogram 3 is 38.8501
    *****************
the mean value of histogram 4 is 1.84617
    *****************
the mean value of histogram 5 is 15.2571
    *****************
the mean value of histogram 6 is 12.5736
    *****************
the mean value of histogram 7 is 15.1297
    *****************
the mean value of histogram 8 is 3.10495
    *****************
the mean value of histogram 9 is 1.57466
    *****************
the mean value of histogram 10 is -0.042204
    *****************
root [1] .q
giuseppemandaglio@Mac book_code %
```

Fig. 5.4 Execution of "MyFirstHist10.C" macro by Root

5.3.1.1 My First Ten Histograms with Root

As an exercise, we show how to repeat the operation in the Sect. 5.3.1, computing
10 histograms using the data from a file sorted in 10 columns. To solve the problem,
we modify the Code 5.5, remembering that there are repetitive actions, we must use
a loop structure.

```
{//Macro "MyFirstHist10.C"
//Please note that the present code works only if the file is
    arranged in 10 columns of real numbers!!!
TH1D *hist[10]; //array of 10 pointer to TH1D
for(int i=0; i<10; i++){
    hist[i] = new TH1D(Form("hist%d",i),"",100,0,0);
    }
ifstream reader("data10.dat");
if(reader.fail())
    {
```

```cpp
10        cout<<"the file does not exist!"<<endl;
11        }
12 double a;
13 //TH1D Object array declaration
14 int counter =0;
15 while(reader>>a)
16     {
17         hist[counter]->Fill(a);
18         counter++;
19         if(counter==10) counter = 0;
20     }
21 for(int i=0; i<10; i++){
22     cout<<"the mean value of histogram "<<i+1<<" is "<<hist[i
       ]->GetMean()<<endl;
23     // cout<<"the standard deviation of histogram "<<i+1<<" is
       "<<hist[i]->GetRMS()<<endl;
24     cout<<"      ******************        "<<endl;
25 }
26 }//Macro end
```

Code 5.5   Root Macro computing ten histograms

How the macro works is shown in Fig. 5.4. If we want a graphical representation
of data, we can use the TCanvas class and the method "Divide" to divide the areas
into sub-canvases of numbers large enough to store all figures. By adding to the
previous code the following instructions reported in Code 5.6:

```cpp
1 //To be added to the previous macro
2 //.....
3
4 TCanvas *box = new TCanvas("box","",800,400);
5 box->Divide(5,2); //ten sub-canvas
6
7 for(int i=0; i<10;i++){
8  box->cd(i+1); //to move into the sub-canvas
9  hist[i]->Draw();
10 }
11 //To save the figures:
12 box->SaveAs("10histo.pdf");
13 }//end of macro
```

Code 5.6   Code snippet showing how to divide a canvas into multiple areas to display additional
figures

and the result is in the following figure: The representation of the ten histograms
shown in Fig. 5.5 is of relatively modest quality. Still, by using the methods offered
by ROOT to control every aspect of the graphical data representation, it is possible
to produce figures of the highest quality and readability.

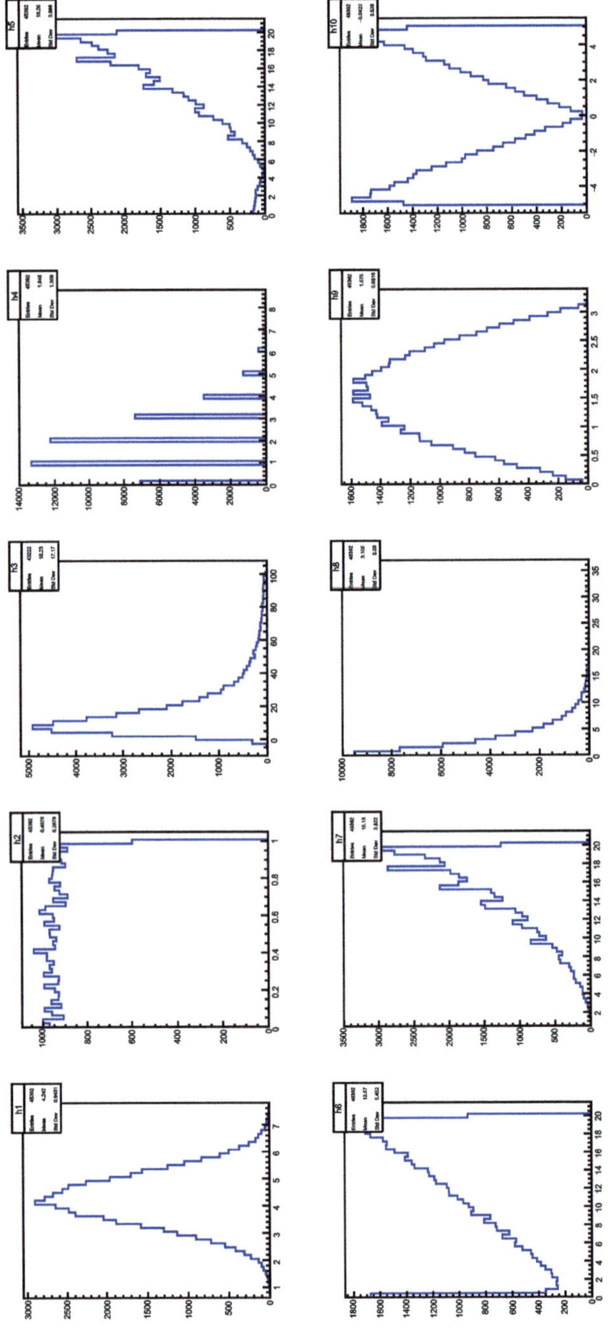

Fig. 5.5 Example of graphical representation of ten histograms. The readability of the text can be improved by using the proper available class methods

5.3.1.2   Figures of High Quality with Root

Aesthetic quality is an important aspect to consider when visualizing data. While clarity and comprehensive labeling are essential, avoiding overcrowding the figure with excessive text is equally important. ROOT makes creating high-quality visualizations well-suited for public presentations, academic theses, journal articles, or professional reports easy. In the code snippet shown in Fig. 5.7, we present a macro that generates a simple yet refined figure, with careful attention to the sizing of axis labels and titles, as discussed in paragraph Sect. 5.2.1:

```
1  {
2  //general settings
3  int mymarkerstyle=20;
4  float mymarkersize=1.;
5  float mytextsize=0.065;
6  float mytextsize1=0.055;
7  int mytextfont=132; //times new roman...
8
9  TCanvas *Box = new TCanvas("Box","Box",600,500); //Constructor
       600pt x 550 pt
10 gStyle->SetOptStat(0); //no Stat information
11 Box->SetFillColor(0);//Background color, 0 means white
12 Box->SetBorderMode(0);//No border
13 Box->SetLeftMargin(0.13); //13% of the largeness of the figure
       as left margin
14 Box->SetRightMargin(0.05);// 5% of the largeness of the figure
       as right margin
15 Box->SetTopMargin(0.07); //7% of the largeness of the figure
       as top margin
16 Box->SetBottomMargin(0.12); //12% of the largeness of the
       figure as bottom margin
17
18 //Just to have data to be preresented
19 TH1D *hist = new TH1D("hist","",100,-5,5);
20 auto func = new TF1("func","x*gaus(0) -x*x+[3]",-5,5);
21 func->SetParameters(100,0,1,1000);
22 hist->FillRandom("func",100000000);
23
24  hist->GetYaxis()->SetTitleSize(mytextsize);
25  hist->GetXaxis()->SetTitleSize(mytextsize);
26  hist->GetXaxis()->SetLabelSize(mytextsize1);
27  hist->GetYaxis()->SetLabelSize(mytextsize1);
28  hist->GetXaxis()->SetTitleFont(mytextfont);
29  hist->GetYaxis()->SetTitleFont(mytextfont);
30  hist->GetXaxis()->SetLabelFont(mytextfont);
31  hist->GetYaxis()->SetLabelFont(mytextfont);
32  hist->GetXaxis()->SetNdivisions(108);
33  hist->GetXaxis()->CenterTitle(1);
34  hist->GetYaxis()->CenterTitle(1);
35  hist->GetXaxis()->SetTitleOffset(0.95);
36  hist->GetYaxis()->SetTitleOffset(0.98);
37  hist->SetTitle("");
```

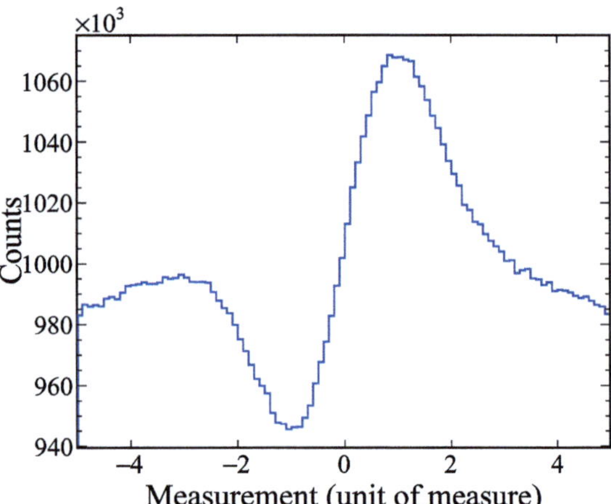

Fig. 5.6 Example of a good quality representation of a histogram

```
38   hist->GetYaxis()->SetTitle("Counts");
39   hist->GetXaxis()->SetTitle("Measurement (unit of measure)");
40   hist->SetLineColor(4);
41   hist->Draw();
42
43   Box->SaveAs("simplefigure.pdf");
44 }//end of macro
```

Code 5.7 Root macro computing an histogram taking care of axis, labels, fonts etc.

and the result produced by the macro is reported in Fig. 5.6.

When dealing with more complex figures that include multiple datasets, adding a legend to clarify the relationship between line styles or markers and the corresponding data can be helpful. ROOT provides the TLegend class specifically for this purpose. In the example shown in Code 5.8, we demonstrate the key features of this class, including how to set its position and size, customize the font and text size, and associate legend entries with the plotted data:

```
1  {
2  //general settings
3  int mymarkerstyle=20;
4  float mymarkersize=1.;
5  float mytextsize=0.065;
6  float mytextsize1=0.055;
7  int mytextfont=132;  //times new roman...
8
9  TCanvas *Box = new TCanvas("Box","Box",600,500); //Constructor
          600pt x 550 pt
10 gStyle->SetOptStat(0); //no Stat information
11 Box->SetFillColor(0);//Background color, 0 means white
12 Box->SetBorderMode(0);//No border
13 Box->SetLeftMargin(0.17); //17% need more space with respect
          to the previous figure
```

```cpp
14  Box->SetRightMargin(0.05);// 5% of the largeness of the figure
        as the right margin
15  Box->SetTopMargin(0.07);  //7% of the largeness of the figure
        as the top margin
16  Box->SetBottomMargin(0.12);  //12% of the largeness of the
        figure as the bottom margin
17
18  //Just to have data to be represented
19  TH1D *hist0 = new TH1D("hist0","",50,0,10);
20  TH1D *hist1 = new TH1D("hist1","",50,0,10);
21  TH1D *hist2 = new TH1D("hist2","",50,0,10);
22  TH1D *hist3 = new TH1D("hist3","",50,0,10);
23
24  auto func = new TF1("func","gaus(0)",0,10);
25  func->SetParameters(100,1,0.5);
26  hist0->FillRandom("func",100000);
27  func->SetParameters(100,2,0.75);
28  hist1->FillRandom("func",100000);
29  func->SetParameters(100,3,1);
30  hist2->FillRandom("func",100000);
31  func->SetParameters(100,4,1.5);
32  hist3->FillRandom("func",100000);
33
34   hist0->GetYaxis()->SetTitleSize(mytextsize);
35   hist0->GetXaxis()->SetTitleSize(mytextsize);
36   hist0->GetXaxis()->SetLabelSize(mytextsize1);
37   hist0->GetYaxis()->SetLabelSize(mytextsize1);
38   hist0->GetXaxis()->SetTitleFont(mytextfont);
39   hist0->GetYaxis()->SetTitleFont(mytextfont);
40   hist0->GetXaxis()->SetLabelFont(mytextfont);
41   hist0->GetYaxis()->SetLabelFont(mytextfont);
42   hist0->GetXaxis()->SetNdivisions(510);
43   hist0->GetXaxis()->CenterTitle(1);
44   hist0->GetYaxis()->CenterTitle(1);
45   hist0->GetXaxis()->SetTitleOffset(0.95);
46   hist0->GetYaxis()->SetTitleOffset(1.3);//move the title
47   hist0->SetTitle("");
48   hist0->GetYaxis()->SetTitle("Counts");
49   hist0->GetXaxis()->SetTitle("Measurement (unit of measure)");
50   hist0->SetLineColor(4);
51   hist0->Draw();
52   hist1->SetLineColor(2);
53   hist1->SetLineStyle(2);
54   hist1->Draw("same");
55   hist2->SetLineColor(4);
56   hist2->SetMarkerColor(4);
57   hist2->SetMarkerStyle(20);
58   hist2->Draw("samepe");
59   hist3->SetLineColor(2);
60   hist3->SetMarkerColor(2);
61   hist3->SetMarkerStyle(24);
62   hist3->Draw("samepe");
63
```

Fig. 5.7 Example of a representation of multiple histograms, with different symbols and a legend

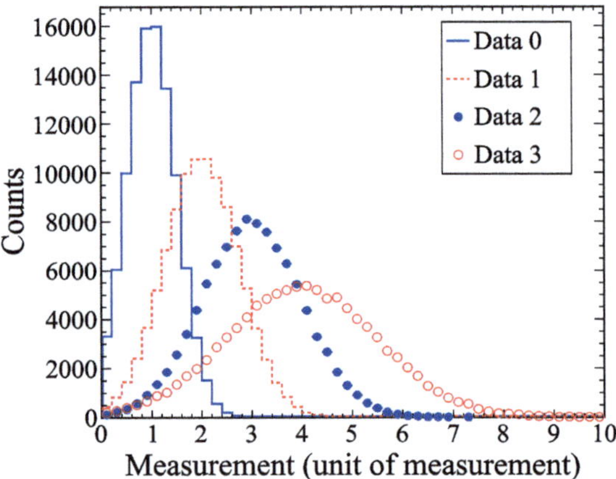


```
64  TLegend *leg = new TLegend(0.7,0.6,0.9,0.9,NULL,"brNDC");
65  //                                    o(0.9,0.9) in % of figure
66  //                   |                |
67  //                   |                |
68  //                   |                |
69  //                   |                |
70  //(0.7,0.6)o--------------------------
71
72     leg->SetBorderSize(1);
73     leg->SetLineColor(1);
74     leg->SetLineStyle(1);
75     leg->SetLineWidth(1);
76     leg->SetFillColor(0);
77     leg->SetTextSize(mytextsize1);
78     leg->SetTextFont(mytextfont);
79     leg->SetFillStyle(1001);
80     //how to link data and symbols
81     leg->AddEntry(hist0,"Data 0","l");
82     leg->AddEntry(hist1,"Data 1","l");
83     leg->AddEntry(hist2,"Data 2","p");
84     leg->AddEntry(hist3,"Data 3","p");
85
86     leg->Draw();
87
88   Box->SaveAs("lesssimplefigure.pdf");
89  }
```

Code 5.8 Root Macro computing histograms and introducing a legend in the figure

As shown in the code, the constructor of the TLegend class requires the coordinates of the legend box's bottom-left and top-right corners within the figure. These coordinates are specified as percentages of the canvas width and height for the x and y axes, respectively. The output of the macro is displayed in Fig. 5.7.

Similarly, data can be visualized using the TGraph and TF1 classes, which share many of the same methods and options as the TH1 class. In the following code example, we demonstrate how different classes can handle various data types while employing consistent methods to display them within the exact figure.

```
 1        {
 2  //general settings
 3  int mymarkerstyle=20;
 4  float mymarkersize=1.;
 5  float mytextsize=0.065;
 6  float mytextsize1=0.055;
 7  int mytextfont=132; //times new roman...
 8
 9  TCanvas *Box = new TCanvas("Box","Box",600,500); //Constructor
        600pt x 550 pt
10  gStyle->SetOptStat(0); //no Stat information
11  Box->SetFillColor(0);//Background color, 0 means white
12  Box->SetBorderMode(0);//No border
13  Box->SetLeftMargin(0.17); //17% need more space
14  Box->SetRightMargin(0.05);// 5% of the largeness of the figure
        as right margin
15  Box->SetTopMargin(0.07); //7% of the largeness of the figure
        as top margin
16  Box->SetBottomMargin(0.12); //12% of the largeness of the
        figure as bottom margin
17
18  //Just to have data to be presented
19  TH1D *hist0 = new TH1D("hist0","",50,0,10);
20
21
22  auto func = new TF1("func","gaus(0)",0,10);
23  func->SetParameters(100,1,0.5);
24  hist0->FillRandom("func",10000);
25
26  TH1D *hist1 = new TH1D("hist1","",50,0,10);
27  func->SetParameters(100,5,1.5);
28  hist1->FillRandom("func",10000);
29  //we change the parameters to represent a different shape
30  //of the curve in the figure
31  func->SetParameters(1100,3,1);
32  //We use the variable object, and not the pointer
33  //as an exercise with this kind of variable
34  TGraphErrors measure("data.dat");
35  //operator point and not arrow to access the method!
36  measure.SetName("measure");
37
38    hist0->GetYaxis()->SetTitleSize(mytextsize);
39    hist0->GetXaxis()->SetTitleSize(mytextsize);
40    hist0->GetXaxis()->SetLabelSize(mytextsize1);
41    hist0->GetYaxis()->SetLabelSize(mytextsize1);
42    hist0->GetXaxis()->SetTitleFont(mytextfont);
43    hist0->GetYaxis()->SetTitleFont(mytextfont);
44    hist0->GetXaxis()->SetLabelFont(mytextfont);
45    hist0->GetYaxis()->SetLabelFont(mytextfont);
46    hist0->GetXaxis()->SetNdivisions(510);
47    hist0->GetXaxis()->CenterTitle(1);
```

```
48  hist0->GetYaxis()->CenterTitle(1);
49  hist0->GetXaxis()->SetTitleOffset(0.95);
50  hist0->GetYaxis()->SetTitleOffset(1.3);//move the title
51  hist0->SetTitle("");
52  hist0->GetYaxis()->SetTitle("Counts");
53  hist0->GetXaxis()->SetTitle("Measurement (unit of measure)");
54  hist0->SetLineColor(4);
55  hist0->Draw();
56  func->SetLineColor(2);
57  func->SetLineStyle(2);
58  func->Draw("same");
59  measure.SetMarkerColor(3);
60  measure.SetLineColor(3);
61  measure.SetMarkerStyle(22);
62  measure.Draw("samep");
63
64
65  TLegend *leg = new TLegend(0.6,0.7,0.9,0.9,NULL,"brNDC");
66    leg->SetBorderSize(1);
67    leg->SetLineColor(1);
68    leg->SetLineStyle(1);
69    leg->SetLineWidth(1);
70    leg->SetFillColor(0);
71    leg->SetTextSize(mytextsize1);
72    leg->SetTextFont(mytextfont);
73    leg->SetFillStyle(1001);
74    leg->AddEntry(hist0,"histogram","l");
75    leg->AddEntry(func,"function","l");
76  //the AddEntry method requires the memory address of the
        object to be linked
77  //with the symbol in the legend
78  //measure is not a pointer, that why we use the operator & to
        pass the memory address
79  //to the method
80    leg->AddEntry(&measure,"measure","p");
81
82    leg->Draw();
83
84  Box->SaveAs("mixlesssimplefigure.pdf");
85  }
```

Code 5.9 Example of the use of TGraph, TF1 classes

The repetition of nearly identical code snippets might appear redundant at first glance, but it highlights the value of code reusability made possible by the shared methods across different classes. This is a significant advantage of object-oriented programming. While creating figures using a graphical interface is straightforward and doesn't require programming knowledge, this approach becomes time-consuming when dealing with complex analyses involving diverse datasets. In contrast, generating figures through macro code requires an initial investment of time to write. However, once implemented, the code can be easily and quickly adapted to accommodate different

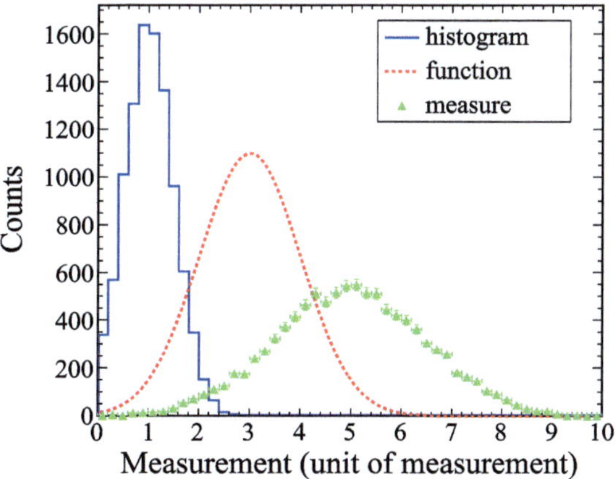

Fig. 5.8 Example of a representation of different kinds of data

data analysis needs. This flexibility is clearly demonstrated in the last three examples. The PDF file produced by the last code is presented in the Fig. 5.8.

In the previous examples, each code concluded by saving the resulting figure as a PDF file, which was then included in this book. The PDF format is widely used for its portability and reliability, particularly because it preserves the content in a non-editable form. However, if we want to modify a figure, we must first update and rerun the code. This raises the question: Can the figure be stored in a format that avoids reprocessing the data? ROOT addresses this need through the TFile class, which allows data and objects to be saved in a binary file format readable by ROOT. This enables us to preserve the results of an analysis and reuse them later without repeating the entire data processing workflow.

5.3.1.3   Store Data on a ROOT TFile

The following section demonstrates how ROOT, through the TFile class, allows data to be stored in a binary file with the .root extension, which is both lightweight and provides fast access. ROOT enables the storage of entire class objects, such as TH1F. The constructor requires the file name to use TFile. If the file already exists and you want to overwrite it, you can add the "RECREATE" option to delete the existing file and create a new one with the same name. Calling the class's Write method is sufficient to save an object to a ROOT file. For instance, the histograms generated in the Code 5.8 can be saved into a ROOT file as follows:

```
{
//macro tfileme.C
//Just to have data to be represented
TH1D *hist0 = new TH1D("hist0","",50,0,10);
```

```cpp
5  TH1D *hist1 = new TH1D("hist1","",50,0,10);
6  TH1D *hist2 = new TH1D("hist2","",50,0,10);
7  TH1D *hist3 = new TH1D("hist3","",50,0,10);
8
9  auto func = new TF1("func","gaus(0)",0,10);
10 func->SetParameters(100,1,0.5);
11 hist0->FillRandom("func",100000);
12 func->SetParameters(100,2,0.75);
13 hist1->FillRandom("func",100000);
14 func->SetParameters(100,3,1);
15 hist2->FillRandom("func",100000);
16 func->SetParameters(100,4,1.5);
17 hist3->FillRandom("func",100000);
18 TFile *saveme = new TFile("histos.root","recreate");
19 hist0->Write();
20 hist1->Write();
21 hist2->Write();
22 hist3->Write();
23 saveme->Close();
24 }
```

Code 5.10 Root Macro computing histograms and saving them in a root file

The macro shown in Code 5.10 computes four histograms, creates a ROOT file named "histos.root", and stores all the histograms within it. ROOT files are lightweight and can be easily read by the ROOT framework. They can be opened either through a macro or the ROOT terminal. To open a ROOT file via the terminal, you can run the following command: *root nameofthefileroot.root*. This launches the ROOT interpreter and opens the specified file. Alternatively, you can start the ROOT interpreter and then open the file using a TFile object *TFile *readme = new TFile("nameoftherootfile.root");* All objects stored in the root file are reusable, as shown in Fig. 5.9.

Suppose we want to produce a high-quality figure using an object stored in a ROOT file. In that case, opening the ROOT file within a macro is sufficient, and then we can customize and generate the figure using the TCanvas class and the methods we've previously learned. In Fig. 5.10, we reused the Code from 5.8, removing the lines that compute the histograms and replacing them with a command that opens the ROOT file. In general, it's often more convenient for complex data analysis workflows to separate the code to perform the data processing and store results from the code producing the final figures. The first type of code focuses on computing and saving results, without considering the quality of the figures. Later, dedicated code can be written to produce high-quality plots that present the key findings.

ROOT files can also store data derived from measurements, simulations, or calculations. For instance, imagine a monitoring system that records environmental variables, such as temperature, humidity, air quality, and pressure, every 10 s, or a simulation that generates many different variables per event. In such cases, one straightforward way to store the data is to use an ASCIIfile, organizing the data so that each row corresponds to a complete measurement or event, and each column represents a specific variable. While this method may be practical for small datasets,

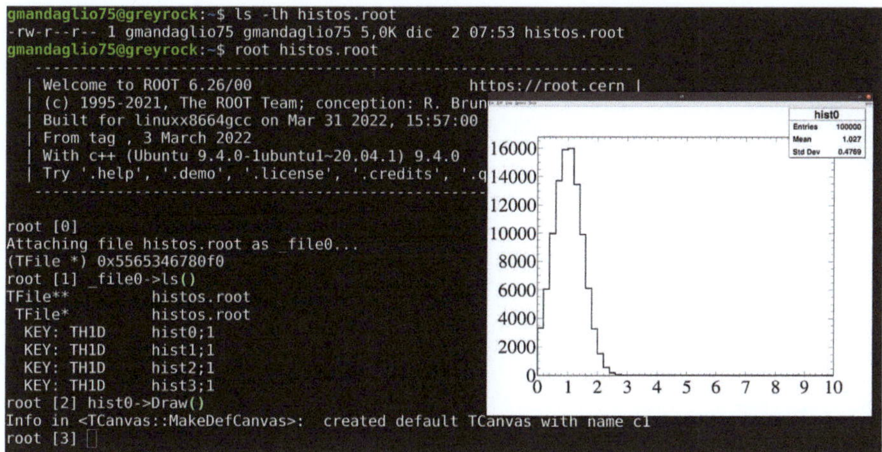

Fig. 5.9 Opening of a root file with the Root interpreter, drawing of a histogram of the TH1D object hist1

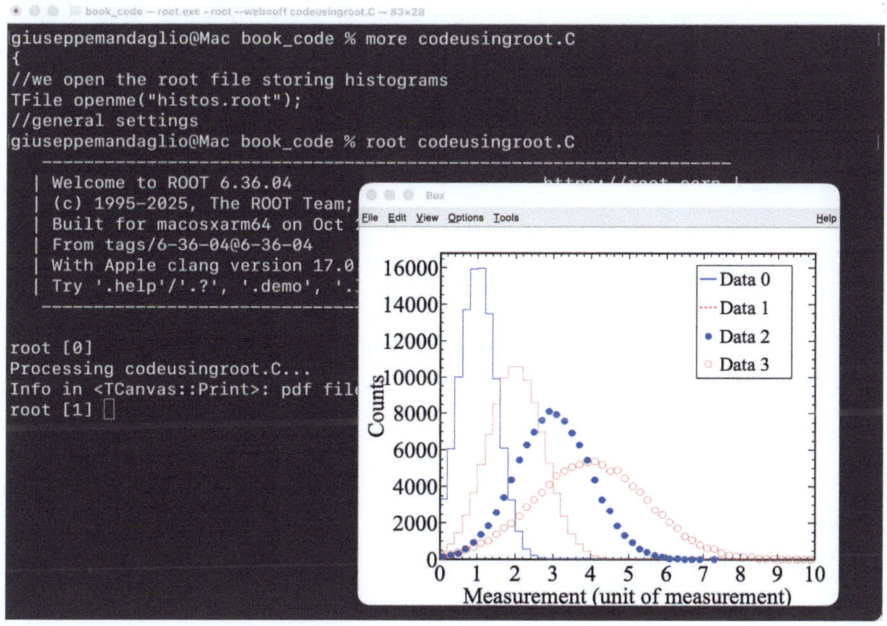

Fig. 5.10 Example of using a root file storing histograms in a macro code

is it still efficient when dealing with large-scale data, millions or even billions of rows? To explore this question, we can perform a simple exercise by writing a code that mimics a measurement process repeated billions of times. We'll use a toy Monte Carlo approach to generate the data, as illustrated in the following example:

```
{
//Class for pseudo-random numbers generator
TRandom3 genme(0);
ofstream writeme("data.dat");
for(int i=0;i<100000000;i++){
writeme<<i<<"   "<<genme.Gaus(0,0.5)<<"   "<<genme.Gaus
(1,1.5)<<"   "<<genme.Gaus(2,2.5)
<<"   "<<genme.Landau(1,1.5)<<"   "<<genme.Uniform()<<endl;
}
writeme.close();
}
```

Code 5.11 Code simulating repetition measurements

The code in 5.11 generates 100 million sets of 5 real numbers and saves them in an ASCIIfile named "data.dat", which occupies approximately 5.4 GB of disk space. By rewriting the code using the TFile class to store the data in a ROOT file and the TTree class to organize the data in a tree structure, we can produce a binary file that is both ROOT-readable and significantly more efficient. This binary ROOT file requires only about 1.7 GB of space, roughly three times smaller than the ASCIIversion, and the code executes faster than the original one.

```
{
//Class for pseudo-random numbers generator
TRandom3 genme(0);
float g1, g2, g3, l, u;
TFile saveme ("data.root","recreate");
TTree *mytree = new TTree("mytree","");
mytree->Branch("g1",&g1,"g1/F");
mytree->Branch("g2",&g2,"g2/F");
mytree->Branch("g3",&g3,"g3/F");
mytree->Branch("l",&l,"l/F");
mytree->Branch("u",&u,"u/F");

for(int i=0;i<100000000;i++){
g1 = genme.Gaus(0,0.5);
g2 = genme.Gaus(1,1.5);
g3 = genme.Gaus(2,2.5);
l  = genme.Landau(1,1.5);
u  = genme.Uniform();
mytree->Fill();
}
mytree->Write();
saveme.Close();
}
```

Code 5.12 Code simulating repetition measurements, arranging the data in a tree structure provided by the TTree class

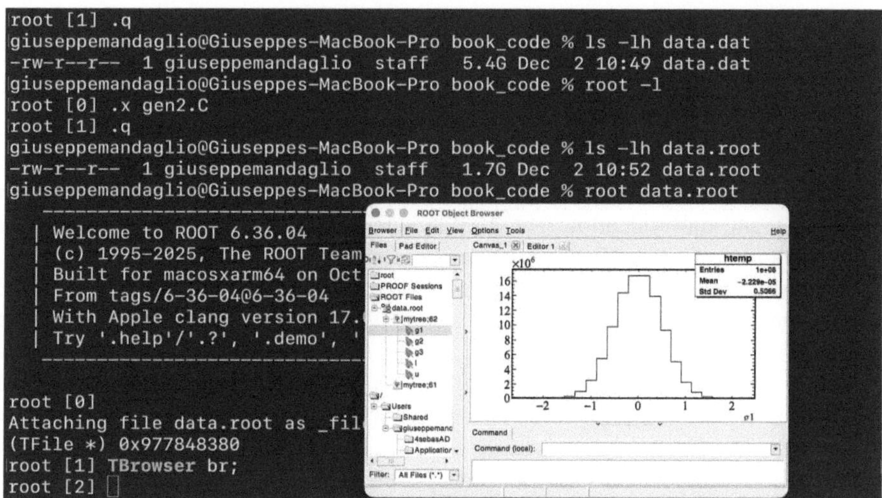

Fig. 5.11 The execution of the Codes 5.11 and Fig. 5.11, the dimension checking of the files produced and the opening of the root file by Root. The file root is explored by using the TBrowser class of Root

The execution of the Codes 5.11 and Fig. 5.11 is presented in the Fig. 5.11.

A ROOT file containing data can be opened directly from the terminal or within a macro, using the `TFile` class. One way is to open the file directly from the terminal by typing:

```
root filename.root
```

In this case, ROOT automatically generates the following line for you:

```
TFile *_file0 = TFile::Open("filename.root");
```

Alternatively, once the ROOT session is active, you can open the file using:

```
TFile openme("filename.root");
```

This command also works when included in a ROOT macro. ROOT provides the `TBrowser` class to visually explore the file's contents, which opens a graphical interface. ROOT will automatically compute and display the corresponding histogram by navigating through the data structure and clicking on one of the leaves (representing stored variables).

5.3.2 *Histograms Algebra*

The TH1 classes offer a wide range of methods, not only for retrieving statistical properties such as the mean, standard deviation, and the bin with the maximum content, but also for performing operations on histograms. In the examples presented below, we will illustrate the use of addition (Add), scaling by a constant (Scale), division (Divide), and multiplication (Multiply). Essentially, any linear operation you might want to perform on histograms can be done using these methods.

The only requirement is that the histograms involved must be defined over the same range and have the same number of bins. It's important to remember that these operations are applied bin-by-bin.

Before exploring these operations in detail, let us take a step back and consider the following exercise: *Two files, daterelli.dat and daterelli2.dat, contain sets of data. The first file has 4 columns, and the second has 2 columns. Load the six datasets from these two files into six histograms, each defined from 0 to 10, with 50 bins.*

We will first show a detailed solution using individual histograms, as presented in Code 5.13.

```cpp
{
    //we use two ifstream variables to read the data
    ifstream read1("daterelli.dat");
    ifstream read2("daterelli2.dat");
    //we reserve 6 histograms as requested
    TH1D *h1 = new TH1D("h1", "", 50, 0, 10);
    TH1D *h2 = new TH1D("h2", "", 50, 0, 10);
    TH1D *h3 = new TH1D("h3", "", 50, 0, 10);
    TH1D *h4 = new TH1D("h4", "", 50, 0, 10);
    TH1D *h5 = new TH1D("h5", "", 50, 0, 10);
    TH1D *h6 = new TH1D("h6", "", 50, 0, 10);
    //we define the maximum number of variables we need for
        reading
    double a, b, c, d;
    //we run two loops for reading the data and computing the
        histograms
    while (read2 >> a >> b) {
        h1->Fill(a);
        h2->Fill(b);
    }
    while (read1 >> a >> b >> c >> d) {
        h3->Fill(a);
        h4->Fill(b);
        h5->Fill(c);
        h6->Fill(d);
    }
    //we close the files after use
    read1.close();
    read2.close();
```

```
28
29  //this is to have them all in the same box (canvas)
30  TCanvas *all = new TCanvas("all", "", 1200, 1000);
31  //we divide the box
32  all->Divide(2, 3);
33  //with cd we move to the sub-sectors of the box
34  all->cd(1);
35  h1->Draw();
36  all->cd(2);
37  h2->Draw();
38  all->cd(3);
39  h3->Draw();
40  all->cd(4);
41  h4->Draw();
42  all->cd(5);
43  h5->Draw();
44  all->cd(6);
45  h6->Draw();
46  }
```

Code 5.13 Histrograms computations from different set of data, long solution

Then we show in Code 5.14, the version of the program using histogram vectors; as you will notice, the code is much smaller and efficient.

```
1  {
2  //we use two ifstream variables to read the data
3  ifstream read1("daterelli.dat");
4  ifstream read2("daterelli2.dat");
5  //we declare a vector of 6 TH1D variable addresses
6  TH1D *histo[6];
7  //we fill the just declared variables with a loop
8  //We use Form to create the internal labels of the histograms
       dynamically
9  //if we define two histograms with the same internal label,
       root gives an error!!!
10  for(int i = 0; i < 6; i++) {
11  histo[i] = new TH1D(Form("h%d", i + 1), "", 50, 0, 10);
12  }
13  //we define the maximum number of variables we need for
       reading
14  double a, b, c, d;
15  //we run two loops for reading the data and computing the
       histograms
16  while (read1 >> a >> b >> c >> d) {
17  histo[2]->Fill(a);
18  histo[3]->Fill(b);
19  histo[4]->Fill(c);
20  histo[5]->Fill(d);
21  }
22  while (read2 >> a >> b) {
23  histo[0]->Fill(a);
24  histo[1]->Fill(b);
25  }
26  //We close the files after use
```

```
27  read1.close();
28  read2.close();
29  //This is to have them all in the same box (canvas)
30  TCanvas *all = new TCanvas("all", "", 1200, 1000);
31  //We divide the box
32  all->Divide(2, 3);
33  //We run a loop to load the figures into the sectors of the
        box
34  for (int i = 0; i < 6; i++) {
35  //with cd, we move to the sub-sectors of the box
36  all->cd(i + 1);
37  histo[i]->Draw();
38  }
39  //We save our figure to a file in PDF format (various formats
        are available: JPG, PNG, eps, PS, etc.)
40  all->SaveAs("allfigures.pdf");
41  //from this point on in the macro, the histograms have been
        computed and are therefore available for our analyses
42  }
```

Code 5.14 Histrograms computations from different set of data, smarter solution

Figure 5.12 displays the computed histograms. The first two histograms in the top panels of the figure are obtained from the file daterelli2.dat (2 columns), while the remaining ones are from the file daterelli.dat (4 columns). By observing these figures, we invite the reader to find the similarities.

Any ideas? ... No. No problem, let's plot our results more readably, for example, by forcing the count scales to be all equal and matching those of the histogram with the maximum value, which in our case is histogram 1 (in this case, 10000).

To adjust the y-axis scale, we added the following code instructions:

```
1  {
2  //to be added to
3  for(int i = 0; i < 6; i++) {
4      tutti->cd(i+1);
5      // we add this instruction inside the loop to force the y-
        axis
6      // to have the same range of variability
7      histo[i]->GetYaxis()->SetRangeUser(0, 10000);
8      histo[i]->Draw();
9  }
10  tutti->SaveAs("figuratutti2.pdf");
11  }
```

The result is shown in Fig. 5.13. While this figure displays the same information as the previous one, it is noticeably easier to interpret. In fact, after a brief observation, it becomes apparent that histogram 1 appears to be the sum of histograms 3 and 5, while histogram 2 seems to be the sum of histograms 4 and 6. This observation is correct. The data files were constructed intentionally in this way: the file with four columns contains the original four-component histograms, whereas the two-column file includes combined data—specifically, the first column is the sum of the data from histograms 3 and 5, and the second column combines data from histograms 4 and 6.

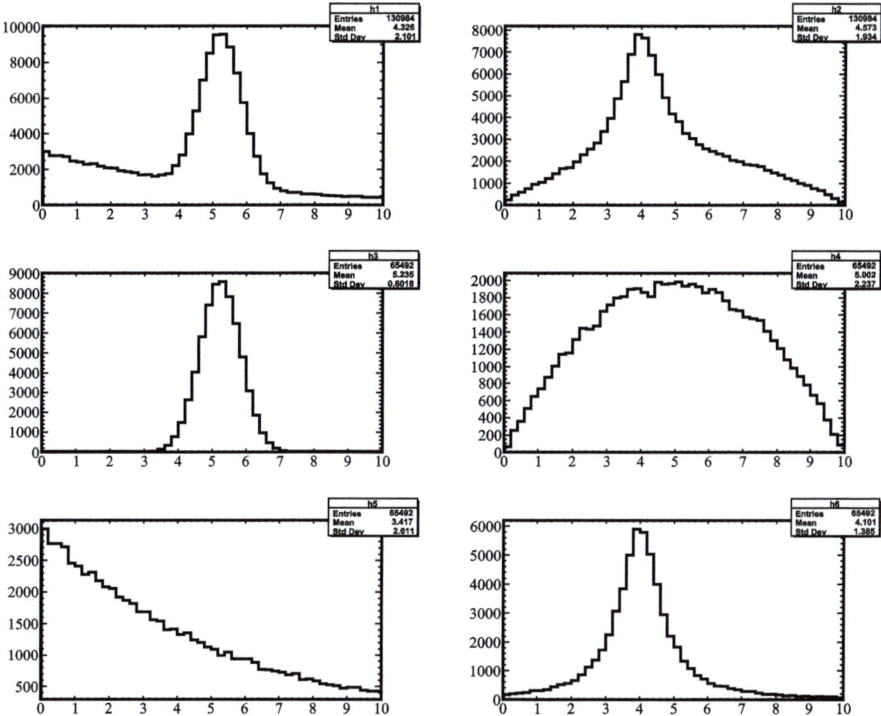

Fig. 5.12 Representation in the same canvas of the 6 histograms loaded from the two files daterelli.dat and daterelli2.dat

These objects we have computed represent an ideal opportunity to test the addition method offered by the TH1D class.

The addition of histograms is performed using the `Add` method. Below are two of its definitions:

```
// These are two of the method's definitions
// The constants, if not defined at the time of use, default
    to 1
Bool_t Add(const TH1* h, const TH1* h2, Double_t c1 = 1,
    Double_t c2 = 1)   // *MENU*
Bool_t Add(const TH1* h1, Double_t c1 = 1)
```

In the first method definition, we observe that it returns a boolean value indicating whether the operation was successful. The input parameters include two pointers to histograms and two multiplicative constants corresponding to each histogram. This method effectively performs a linear combination of the form:

$$h3 = c1 \cdot h1 + c2 \cdot h2$$

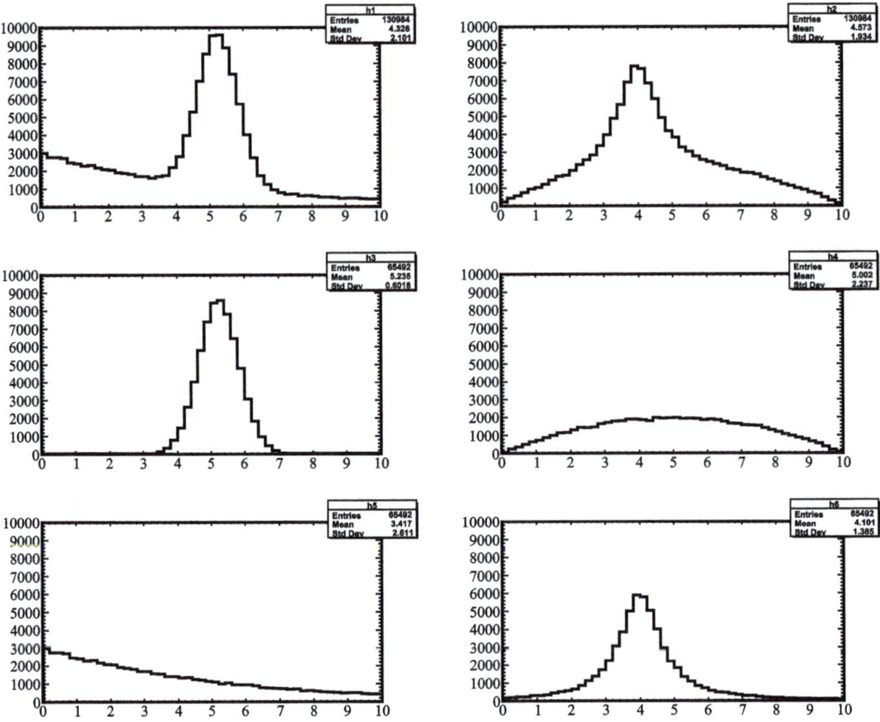

Fig. 5.13 Representation in the same canvas of the six histograms loaded from the two files daterelli.dat and daterelli2.dat, with the same variability range for the y-axis

From this, we immediately see that the method allows us to perform weighted additions and subtractions of histograms. The second method enables us to add to the histogram by invoking the method, using the following syntax:

$$h2 = h2 + c1 \cdot h1$$

As an exercise, let's plot histogram `h1` and overlay it with the difference between `h1` and `h3`, and observe the result:

```cpp
// Assuming we are at the end of the code that computed the
    histograms
// We define a histogram in the same way as the histograms
    that will be summed
TH1D *sum = new TH1D("sum", "", 50, 0, 10);
// We use the Add method, but note that the coefficient c2 in
    this case is -1
sum->Add(h1, h3, 1, -1); // c2 = -1 defines a subtraction
// We use the SetLine... methods to make the figures more
    visible
h1->SetLineColor(1);
```

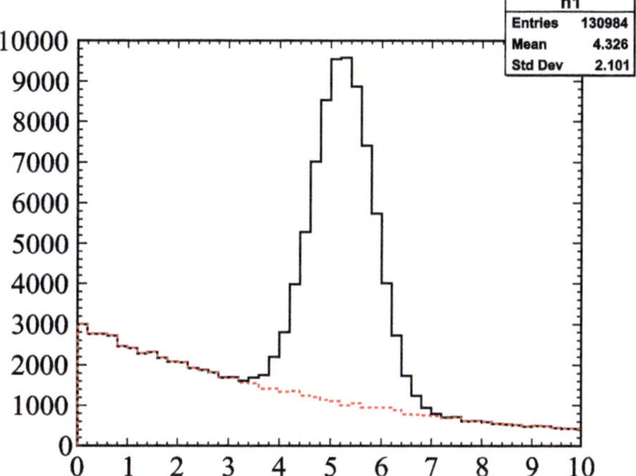

Fig. 5.14 The solid line represents histogram h1, while the dashed line is the result of h1–h3

```
 8  h1->SetLineStyle(1);
 9  h1->SetLineWidth(2);
10  sum->SetLineColor(2);
11  sum->SetLineStyle(2);
12  sum->SetLineWidth(2);
13  // We create a canvas
14  TCanvas *sss = new TCanvas("sss", "", 1200, 1000);
15  // We draw the source graph
16  h1->Draw();
17  // We draw the graph resulting from our subtraction
18  sum->Draw("same"); // the "same" option overlays the plot
        instead of replacing it
19  // We save the file to include in this manual
20  sss->SaveAs("diffhisto1.pdf");
```

As shown in Fig. 5.14, the dashed curve, obtained by subtracting histogram h3 from h1, perfectly matches histogram h5. If you're skeptical (which is always a good attitude in data analysis), you can perform an additional check by overlaying the histogram we named sum (h1-h3) with histogram h5 to verify that they indeed coincide. **Note:** when doing this, be sure to change the line styles or colors of the curves. Since the histograms are identical, without distinguishing styles, it may appear that one of the curves has disappeared due to complete overlap.

A similarly valuable exercise is to repeat this verification using histograms h2, h4, and h6. In fact, in the left panel of Fig. 5.15, we show histogram h2 (blue line) and the difference h2-h4 (red line), which yields histogram h6. In the right panel of the exact figure, we display histogram h2 (blue line) and the difference h2-h6 (red line), recovering histogram h4.

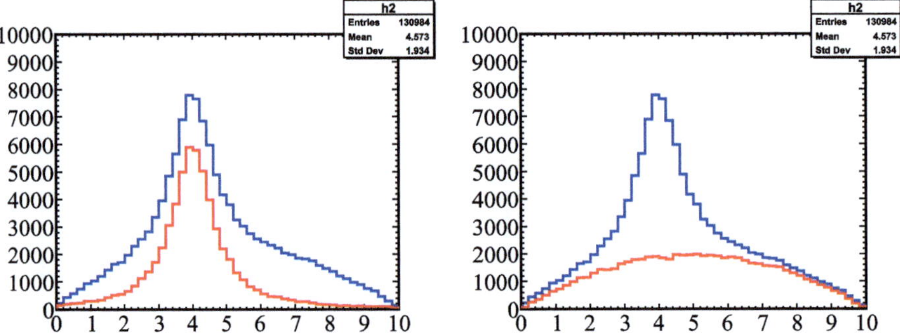

Fig. 5.15 The blue line represents histogram h2, while the red line is the result of h2–h4 on the left and h2–h6 on the right

5.3.3 Data Modelling

The TF1 class enables the creation of objects that represent a function defined over a specified interval, tailored to the user's needs. In this section, we will use the simplest form of its constructor, whose syntax is shown below:

```cpp
// Constructor
TF1(const char* name, const char* formula, Double_t xmin = 0,
    Double_t xmax = 1, TF1::EAddToList addToGlobList =
    EAddToList::kDefault)
// Example of a non-parametric function; we omitted the last
    option, and it assumes the default.
TF1 *function1 = new TF1("function1", "sin(x)/x", -10, 10);
// Example of a parametric function
TF1 *function2 = new TF1("function2", "[0]*exp(-pow(x-[1],2)
    /[2])", 0, 7.4);
// In this case, we have 3 parameters: [0], [1], [2]
```

Functions created using the TF1 class can be plotted to generate a sequence of values or used as parametric models for data fitting. As we will see later, they can also serve to define data distributions that follow the probability density function they represent.

As an exercise, we will use the histograms shown in Fig. 5.12 to address the following questions: Can the trends observed in these histograms be described by a mathematical function? Do the parameters of such a function carry any physical meaning? Can they be considered measurable quantities?

Among the shapes shown in Fig. 5.12, histograms 3 and 4 appear to be the simplest. The third histogram resembles a Gaussian distribution, while the fourth appears to follow a parabolic curve, or more generally, a second-degree polynomial.

ROOT provides built-in support for common functional forms such as Gaussians, polynomials (up to 9th degree), and others, using predefined keywords. To extract the parameters of the Gaussian function that may fit the third histogram's distribution, we can use the code shown in Listing 5.15:

```cpp
{
  ifstream readfile("daterelli2.dat");
  // declare an array of 4 pointers to TH1D variables
  TH1D *histo[4];
  // use a loop to fill the just declared variables
  for(int i = 0; i < 4; i++){
      histo[i] = new TH1D(Form("h%d", i + 1), "", 50, 0, 10);
  }
  // define the maximum number of variables needed for reading
  double a, b, c, d;
  // execute two loops for reading the data and computing the
      histograms
  while(readfile >> a >> b >> c >> d)
  {
      histo[0]->Fill(a);
      histo[1]->Fill(b);
      histo[2]->Fill(c);
      histo[3]->Fill(d);
  }
  // close the file after usage
  readfile.close();

  TCanvas *box = new TCanvas("box", "", 600, 500);

  histo[2]->Draw();

  //Use the keyword gaus directly to
  //Perform a fit with a Gaussian without using a TF1
  histo[0]->Fit("gaus");
  box->SaveAs("fitgaus.pdf");
}
```

Code 5.15   Example of Gaussian fit with ROOT

The result obtained by running the small code we just wrote in ROOT is shown in Fig. 5.16.

As shown in the screenshot in Fig. 5.16, after computing the histogram and displaying it using the Draw() method, the code applies the Fit method with the keyword gaus to superimpose a red Gaussian curve on the histogram plot. In this case, the fit matches the data very well. The ROOT interpreter also prints a set of useful statistical details to the terminal: FNC provides the fit's χ^2, indicates the minimization algorithm used (MIGRAD), the number of iterations performed before convergence, and finally, a list of the fit parameters along with their corresponding uncertainties.

These values can also be accessed programmatically using the fit function's methods: GetParameter(index) retrieves a specific parameter (where the index starts from zero), GetParameters(pointer to a real-valued vector of appropriate size) retrieves all parameters at once, GetChisquare() returns the χ^2 value, and GetNDF() returns the number of degrees of freedom.[1]

---

[1] Knowing the degrees of freedom is important, particularly because it allows you to compute the reduced χ^2, defined as the χ^2 divided by the number of degrees of freedom.

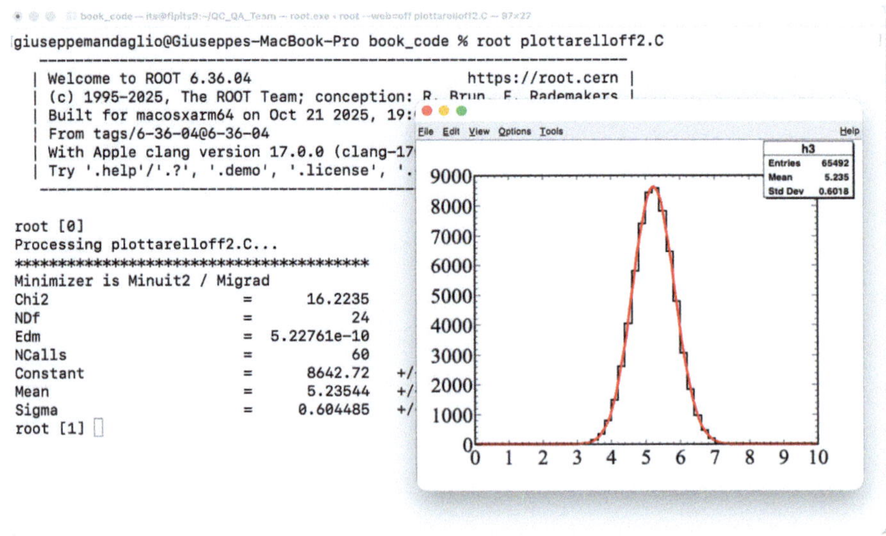

Fig. 5.16 Screenshot showing the execution of the code that produces the Gaussian fit

```
{
......

......

Continuing from the previous code

......

......
TCanvas *box = new TCanvas("box", "", 600, 500);

histo[2]->Draw();

//Use the keyword gaus directly to
// perform a fit with a Gaussian without using a TF1
histo[0]->Fit("gaus");
box->SaveAs("fitgaus.pdf");
cout << "The chi2 of the fit is " << gaus->GetChisquare() <<
    endl;
//Parameter 0 is the amplitude of the Gaussian
cout << "The mean value of the Gaussian is " << gaus->
    GetParameter(1) << endl;
cout << "The standard deviation of the Gaussian is " << gaus->
    GetParameter(2) << endl;
}
```

We could have implemented the Gaussian function independently:

```
{
double low = 0., high = 100.; // the bounds of the function's
    definition interval
TF1 *funz = new TF1("funz", "[0]*exp(-pow(x-[1],2)/[2])", low,
    high);
}
```

Fig. 5.17 On the left is histogram No. 5 from Fig. 5.12 fitted with a straight line; on the right is histogram No. 6 fitted with a Gaussian

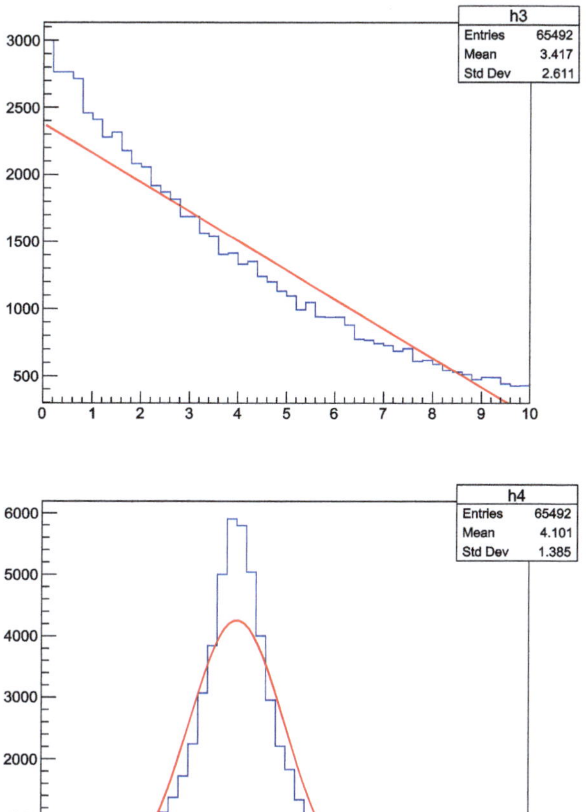

Equally simple is the fit of histogram hist[1]; this has a parabolic trend, and using the keyword *pol2* as an argument of the Fit method immediately returns the correct fit. I leave this simple exercise to you.

In this simple example, we did not encounter particular problems; the distribution was produced with a toy Monte Carlo that generated a Gaussian. You can observe the same thing by repeating the exercise for the second histogram using a parabola. So in these two examples, we had no difficulty achieving the fits, but we will see later that it is not always so simple.

Now let's try to model the last two histograms in Fig. 5.12 with a fit. For the fifth histogram, we try a straight line, while for the sixth, we insist on a Gaussian given the peak shape of the histogram under examination. The result of our attempt is shown in Fig. 5.17.

Qualitatively observing Fig. 5.17, without resorting to χ^2 or other statistical tests, we immediately notice that something is wrong; the models we chose do not work because they do not help represent the data. Let's try to repeat the exercise, using a second-degree polynomial in the first case, and a Breit-Wigner function in the second case:

$$f(x) = A \cdot \frac{\Gamma^2/4}{(x - x_0)^2 + \Gamma^2/4}$$

Parametrized as follows:

```
{
double low = 0., high = 10.; // the bounds of the function's
    definition interval
TF1 *funz = new TF1("funz", "[0]/(pow(x-[1],2)+[2])", low,
    high);
// where parameter [0] represents the amplitude, [1] the peak
    point, [2] the half-width.
```

by using the codes:

```
{
histo[2]->Fit("pol2"); // we use the keyword pol2 for the
    second-degree polynomial
c1->SaveAs("fig1.pdf");
TF1 *funz = new TF1("funz", "[0]/(pow(x-[1],2)+[2])", 0, 10);
histo[3]->Fit("funz", "R"); // we use our parameterization
c1->SaveAs("fig2.pdf");
}
```

We obtain the Fig. 5.18.

In Fig. 5.18, we can observe that the second-degree polynomial provides a good fit to the histogram on the top panel, while it clearly fails to model the data in the histogram on the bottom.

However, the model used for the top panel fit is incorrect. The data were generated from a decaying exponential function. The apparent agreement between the model and the data is purely coincidental—mathematically, a second-degree polynomial can approximate an exponential curve reasonably well over a limited range. This approximation masked the fact that the chosen model was not the correct one.

In contrast, the model is appropriate for the bottom panel's histogram. What went wrong here is not the model itself but rather the behavior of the minimization algorithm used in the Fit process. These algorithms attempt to adjust the model parameters to best match the data, but their search does not continue indefinitely. It is possible for the fitting process to converge to a local minimum of the error function rather than the desired global minimum, which is precisely what happened in this case.

To resolve such issues, it is helpful to provide the fitting function parameters close to the expected solution.

Let's try fitting the data again, this time using the SetParameters(first parameter, second parameter, ...) method to suggest starting values that are near the correct ones for the fit:

Fig. 5.18 On the left is histogram No. 5 from Fig. 5.12 fitted with a parabola; on the right is histogram No. 6 fitted with a Breit-Wigner

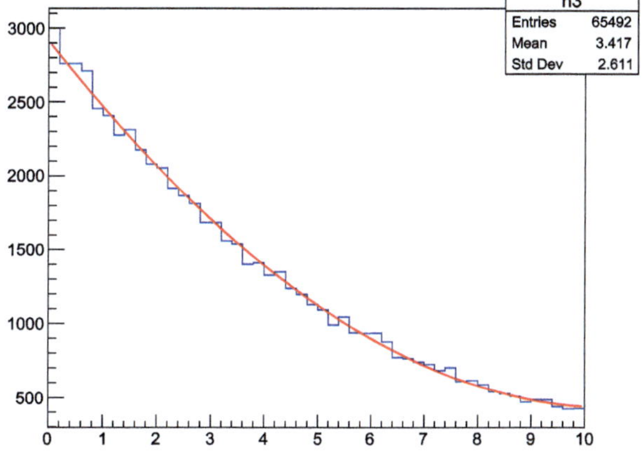

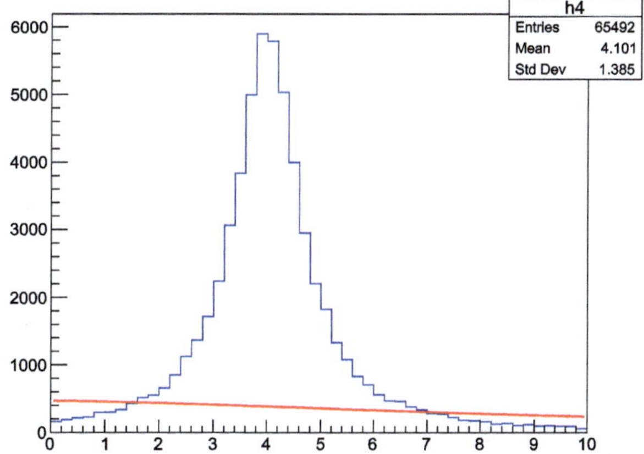

```
{
// box for the figures
TCanvas *box = new TCanvas("box", "", 600, 500);
// define the Breit-Wigner
TF1 *funz = new TF1("funz", "[0]/(pow(x-[1],2)+[2])", 0, 10);
// suggest the initial parameters for the fit
funz->SetParameters(2000, 4, 0.5);
histo[3]->Fit("funz", "R");
box->SaveAs("fitebwsgiusta.pdf");
// repeat the same procedure for the exponential
TF1 *funz2 = new TF1("funz2", "[0]*exp(-[1]*x)", 0, 10);
funz2->SetParameters(1, 1);
histo[2]->Fit("funz2", "R");
box->SaveAs("fitexpgiusta.pdf");
}
```

The result of the correct codes is shown in Fig. 5.19.

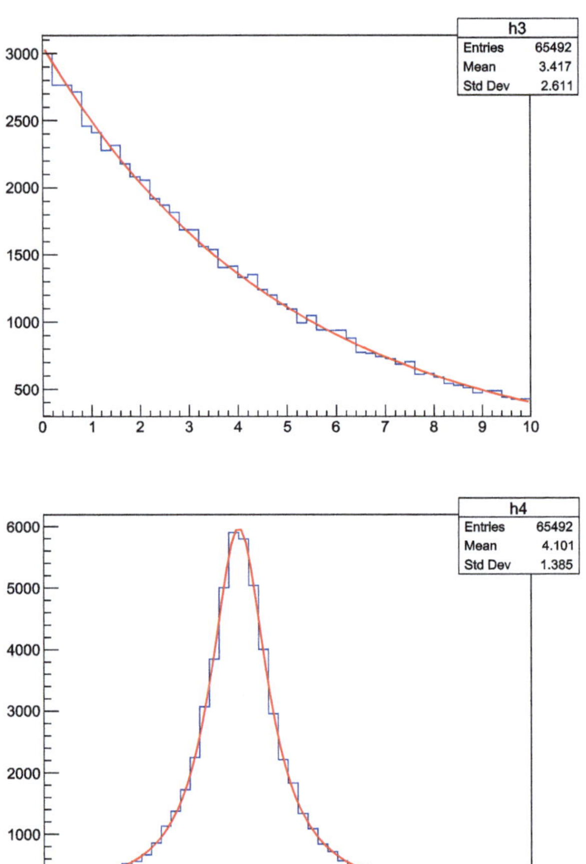

Fig. 5.19 On the top is histogram 5 from Fig. 5.12 fitted with a decreasing exponential; on the bottom is histogram No. 6 fitted with a Breit-Wigner. In both fits, parameters were suggested before performing the fit

As shown in Fig. 5.19, both histograms are accurately described by the proposed models.

This is an important consideration when comparing the top panels of Figs. 5.18 and 5.19. The data fit well with a quadratic polynomial and a decaying exponential in both cases. So, which model should we trust? Both fits are mathematically valid; the key difference lies in the interpretation and the kind of information we aim to extract from the data.

For example, if we know that the data represent the behavior of a population of radioactive nuclides undergoing decay, then the exponential model is more appropriate. It fits the data well and allows us to derive meaningful physical parameters, such as the decay constant.

In conclusion, the appropriateness of a model depends not just on how well it fits the data, but on the context and the underlying nature of the phenomenon being studied. The model you choose should reflect the data's structure and the specific insights you aim to gain from it.

5.3.4   Fit of Histograms with a Composite Shape

In this section, we will attempt to fit a dataset showing an overlap of two signals. The simplest examples with which we can start might be the first two figures shown in Fig. 5.12. In these two cases, we already know the overlap of a Gaussian and an exponential in the first histogram, while we have a parabola and a Breit-Wigner in the second.

Below, we provide a code capable of successfully performing the two fits:

```cpp
{
    ifstream leggimi("daterelli2.dat");

    TH1D *hist[2];
    for(int i = 0; i < 2; i++) {
        hist[i] = new TH1D(Form("h%d", i), "", 50, 0, 10);
    }
    double a, b;
    while(sc >> a >> b) {
        hist[0]->Fill(a);
        hist[1]->Fill(b);
    }
    leggimi.close();

    TCanvas *grafici = new TCanvas("grafici", "", 1200, 600);
    grafici->Divide(2, 1);

    // Fit of histogram 1 composed of exponential + Gaussian
    TF1 *funz1 = new TF1("funz1", "[0]*exp(-[1]*x)", 0, 3);
    funz1->SetParameters(1, 1);
    funz1->SetParNames("a", "b"); // Naming the parameters to
    avoid confusion in the definition of the sum function
    TF1 *funz2 = new TF1("funz2", "[0]*exp(-pow(x-[1],2)/[2])"
    , 3.7, 6.3);
    funz2->SetParameters(1, 1, 1);
    funz2->SetParNames("c", "d", "e");
    funz2->SetLineColor(3);

    grafici->cd(1);

    hist[0]->Draw();
    hist[0]->Fit("funz1", "R"); // R means the fit is limited
    to the data within the function's definition interval
    hist[0]->Fit("funz2", "R+"); // The + option allows
    overlapping the second fit in the figure
```

As we can see in both examples, the data were first fitted using the individual component functions, each applied within the region where it dominates. This strat-

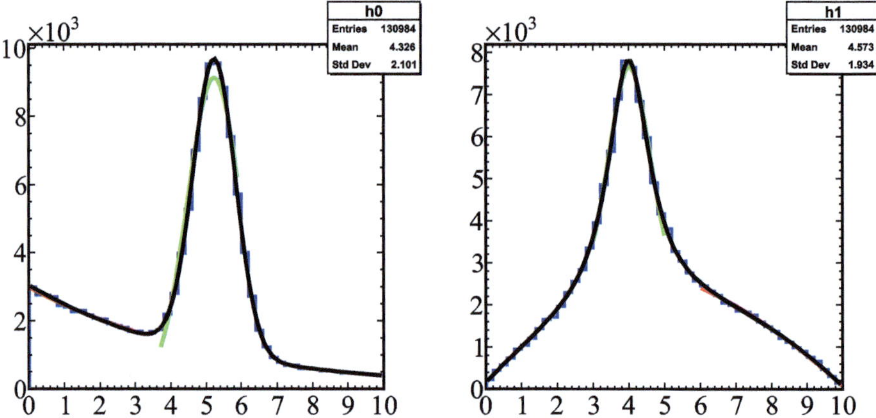

Fig. 5.20 In the right panel of the figure, the fit of histogram hist[0] is shown, while in the left panel is histogram hist[0]. The blue curves represent the histograms, the green and red curves represent the component fits, and the black curves represent the overall fit

egy ensures that the parameters obtained from these preliminary fits are as close as possible to those of the final composite function. Once these initial fits are complete, a sum function is defined over the full data range (Fig. 5.20).

By carefully assigning distinct parameter names to each component function, we avoid conflicts within the sum function. As a result, the sum function automatically uses the parameters from the preliminary fits as initial values for the global fit. The outcome is an excellent overall fit to the data using the composite function. In the figure, the individual component fits are shown in green and red, while the total fit from the sum function is displayed in black.

After the global fit has been successfully performed, the sum function contains the correct parameter values for each component. However, since the component functions were initially defined over restricted subranges of the histogram, we must extend their ranges to span the entire data domain if we want to visualize the full global fit alongside the individual contributions. Additionally, we need a method to transfer the fitted parameters from the sum function back to the component functions. This is illustrated in Code Snippet 5.16:

```
{
  ifstream sc("daterelli2.dat");
  TH1D *hist=new TH1D("hist","",50,0,10);
  double a,b;
  while(sc>>a>>b)
  {
    hist->Fill(b);
  }

  // Fit of histogram 6 composed of Breit-Wigner and parabola
```

```cpp
11   TF1 * fun1= new TF1("fun1","[0]/(pow(x-[1],2)+pow([2]/2,2))",
        3,5);        //((x-[2])*(x-[3])+[1]/4)[0]) alternative form
        of Breit-Wigner
12    fun1->SetParameters(1,1,1,1);
13    fun1->SetParNames("f","g","h","i");
14    fun1->SetLineColor(3);
15    TF1 * fun2= new TF1("fun2","[0]*x*x+[1]*x+[2]", 6,10);
16    fun2->SetParameters(1,1,1);
17    fun2->SetParNames("l","m","n");
18
19    TCanvas * pippo = new TCanvas("pippo","",600,500);
20
21    hist->Draw();
22  // Clone the working histogram to keep a clean copy
23  // not tainted by the fitting processes (just a graphical
        refinement)
24  TH1D *pino = (TH1D*)hist->Clone();
25  // Perform the fits for the component functions
26  hist->Fit("fun1","R");
27  hist->Fit("fun2","R+");
28  // IT IS ESSENTIAL to define the sum after fitting with fun1
        and fun2
29  // sum automatically inherits the parameters from the
        preliminary fits
30  TF1 * sum = new TF1("sum","fun1+fun2", 0, 10);
31  sum->SetLineColor(1);
32  // Global fit with the sum function
33  hist->Fit("sum","R+");
34  // Container to copy parameters after the global fit
35  double parameters[6];
36  // Method to copy all parameters in one go
37  sum->GetParameters(parameters);
38  //sum->GetParameters(&parameters[0]); // equivalent command,
        specifying from which element to start copying
39  // VERY IMPORTANT -> Correct the definition range of the
        component functions
40  fun1->SetRange(0,10);
41  fun2->SetRange(0,10);
42  // Pass the parameters obtained from the global fit to the
        components
43  fun1->SetParameters(&parameters[0]);
44  fun2->SetParameters(&parameters[3]);
45  pino->Draw();
46  // Overlay the clean graph with the sum and component
        functions
47  sum->Draw("same");
48  fun1->Draw("same");
49  fun2->Draw("same");
50  pippo->SaveAs("fitsecondo.pdf");
51  }
```

Code 5.16 Example of determining component functions in a composite–shape fit using ROOT

The result of this code is shown in Fig. 5.21.

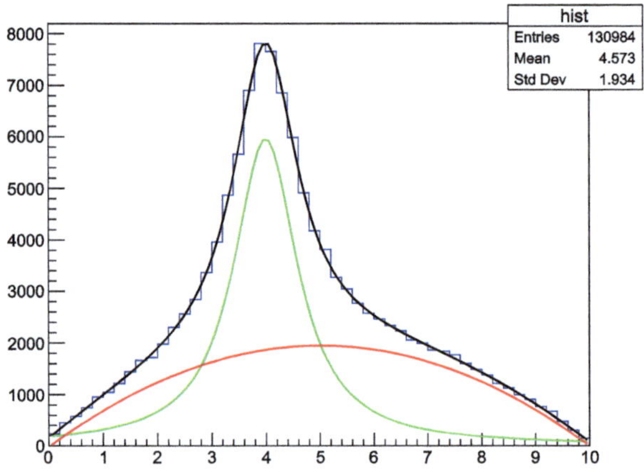

Fig. 5.21 The histogram is shown in blue, the global fit in black, and the component functions with parameters derived from the global fit are shown in red and green

5.3.5   Background Recognition and Subtraction

We learned that data modeling works in two steps. First, we have to choose the functions that best adapt to the data, and second, we have to optimize the parameters of the functions. The optimization of parameters, also in the case of the best choice of the functions, is not a trivial task. If the minimizer algorithm does not receive the parameters initialized to values close to the correct ones, it cannot converge to the best-optimized values of the parameters. In the following code, we represent a case of different signals overlapped on a background.

We will present the code that grows during the analysis process. The first action is plotting data and looking at the results to hypothesize what functions must be used in the fit procedure, as reported in the Code 5.17.

```
1  {
2      TCanvas *window = new TCanvas("window", "window", 600,
       500);
3      ifstream dati1x("dati1x.dat");
4      TH1D *histogram = new TH1D("Histogram", "", 100, 0, 20);
5      double a;
6      while (dati1x >> a) {
7          histogram->Fill(a);
8      }
9      histogram->Draw();
10     window->SaveAs("fiststep.pdf");
11 }
```

Code 5.17 Code simulating repetition measurements, arranging the data in a tree structure provided by the TTree class

By examining Fig. 5.22, it is evident that the spectrum consists of three overlapping peaks superimposed on a background. The peaks can be modeled using either Gaussian or Breit-Wigner functions, while the background appears to follow an exponential trend. But how do we determine whether a given peak is better described by

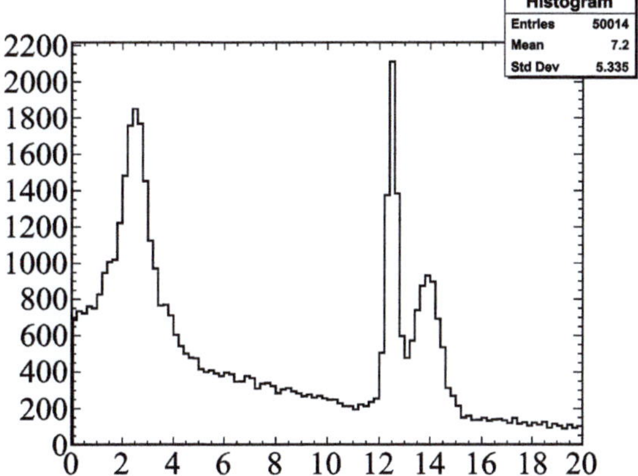

Fig. 5.22 The histogram with three signals and a background

a Gaussian or a Breit-Wigner profile? The answer lies in testing different hypotheses and evaluating which provides the best fit to the data.

In our case, this analysis has already been performed, and we have determined that the first and third peaks are best described by Breit-Wigner functions, while the central peak is a Gaussian. However, we encourage the reader to explore alternative hypotheses and verify whether this configuration indeed yields the best fit.

In the following code, we create two clones of the original histogram (generated by the computer class). The reason for creating these clones will become clear shortly. We then define the individual functions that describe each contribution, both the peaks and the background.

Each function is fitted over the interval where its contribution is dominant, minimizing interference from the others. This allows us to obtain accurate initial parameter estimates, which are then used in a global fit involving a sum of all the component functions.

```cpp
{
    TCanvas *window = new TCanvas("window", "window", 600,
    500);
    ifstream dati1x("dati1x.dat");
    TH1D *histogram = new TH1D("Histogram", "", 100, 0, 20);
    double a;
    while (dati1x >> a) {
        histogram->Fill(a);
    }
    histogram->Draw();
//    window->SaveAs("fiststep.pdf");
    // Create a copy of the original histogram
    TH1D *copy1 = (TH1D*)histogram->Clone();
    TH1D *copy2 = (TH1D*)histogram->Clone();
    // Define the component functions

    TF1 *exponential = new TF1("exponential", "[0]*exp(-x/[1])
    ", 4, 10);
```

```cpp
17   exponential->SetParameters(1, 1);
18   exponential->SetParNames("Aparameter", "Bparameter");
19   exponential->SetLineColor(2);
20   exponential->SetLineStyle(2);
21
22   TF1 *breit_wigner = new TF1("breit_wigner", "[0]/(pow(x
     -[1],2)+pow([2]/2, 2))", 1, 3.5);
23   breit_wigner->SetParameters(1, 1, 1);
24   breit_wigner->SetParNames("Cparameter", "Dparameter", "
     Eparameter");
25   breit_wigner->SetLineColor(4);
26
27   TF1 *breit_wigner2 = new TF1("breit_wigner2", "gaus", 13,
     15);
28   breit_wigner2->SetParameters(1, 1, 1);
29   breit_wigner2->SetParNames("Fparameter", "Gparameter", "
     Hparameter");
30   breit_wigner2->SetLineColor(5);
31
32   TF1 *gaussian = new TF1("gaussian", "gaus", 11.5, 13);
33   gaussian->SetParameters(1, 1, 1);
34   gaussian->SetParNames("Iparameter", "Lparameter", "
     Mparameter");
35   gaussian->SetLineColor(3);
36
37   histogram->Fit("exponential", "R");
38   histogram->Fit("breit_wigner", "R+");
39   histogram->Fit("breit_wigner2", "R+");
40   histogram->Fit("gaussian", "R+");
41   // After fitting the components, define the sum function
42   TF1 *sum_function = new TF1("sum_function", "exponential +
      breit_wigner + breit_wigner2 + gaussian", 0, 20);
43   sum_function->SetLineColor(1);
44   sum_function->SetLineStyle(3);
45
46   histogram->Fit("sum_function", "R+");
47   // Extract fundamental statistical parameters from the sum
      function
48   double ndf = sum_function->GetNDF();
49   double chi_square = sum_function->GetChisquare();
50   double prob = sum_function->GetProb();
51
52   cout << "Degrees of freedom: " << ndf << " Chi-Square: "
     << chi_square << " Prob: " << prob << endl;
53   cout << "The reduced chi-square is: " << chi_square / ndf
     << endl;
54   window->SaveAs("secondstep.pdf");
55
56 }
```

Code 5.18 Code simulating repetition measurements, arranging the data in a tree structure provided by the TTree class

```
root [0]
Processing sd.C...
 FCN=45.4231 FROM MIGRAD      STATUS=CONVERGED      198 CALLS          199 TOTAL
                     EDM=5.1775e-07      STRATEGY= 1       ERROR MATRIX ACCURATE
  EXT PARAMETER                                    STEP         FIRST
  NO.   NAME        VALUE          ERROR          SIZE        DERIVATIVE
   1   Aparameter   8.62934e+02    3.42046e+01    2.75038e-02  -4.97917e-05
   2   Bparameter   7.87128e+00    3.59799e-01    2.89227e-04  -2.00662e-03
 FCN=136.171 FROM MIGRAD      STATUS=CONVERGED      346 CALLS          347 TOTAL
                     EDM=5.96661e-08     STRATEGY= 1       ERROR MATRIX ACCURATE
  EXT PARAMETER                                    STEP         FIRST
  NO.   NAME        VALUE          ERROR          SIZE        DERIVATIVE
   1   Cparameter   1.98528e+03    8.41918e+01    9.18308e-02   1.92276e-05
   2   Dparameter   2.42995e+00    1.19890e-02    6.69790e-05  -1.06170e-02
   3   Eparameter   2.15468e+00    5.61310e-02    6.14955e-05  -3.03455e-02
 FCN=27.6873 FROM MIGRAD      STATUS=CONVERGED       77 CALLS           78 TOTAL
                     EDM=2.63525e-07     STRATEGY= 1       ERROR MATRIX ACCURATE
  EXT PARAMETER                                    STEP         FIRST
  NO.   NAME        VALUE          ERROR          SIZE        DERIVATIVE
   1   Fparameter   9.05875e+02    1.58270e+01    2.99139e-02  -6.26740e-05
   2   Gparameter   1.38566e+01    1.10675e-02    2.67078e-05   2.17387e-03
   3   Hparameter   6.26436e-01    1.31338e-02    1.44297e-05  -7.85091e-02
 FCN=496.095 FROM MIGRAD      STATUS=CONVERGED       86 CALLS           87 TOTAL
                     EDM=5.83341e-11     STRATEGY= 1       ERROR MATRIX ACCURATE
  EXT PARAMETER                                    STEP         FIRST
  NO.   NAME        VALUE          ERROR          SIZE        DERIVATIVE
   1   Iparameter   1.85978e+03    3.61291e+01    2.57522e-01  -1.61316e-07
   2   Lparameter   1.24994e+01    3.77854e-03    4.08490e-05   2.39642e-03
   3   Mparameter   2.73553e-01    4.68681e-03    3.69184e-05  -1.98152e-03
 FCN=113.926 FROM MIGRAD      STATUS=CONVERGED      617 CALLS          618 TOTAL
                     EDM=2.59767e-08     STRATEGY= 1  ERROR MATRIX UNCERTAINTY   4.4 per cent
  EXT PARAMETER                                    STEP         FIRST
  NO.   NAME        VALUE          ERROR          SIZE        DERIVATIVE
   1   Aparameter   6.20842e+02    1.33190e+01   -1.16314e-02  -5.34267e-07
   2   Bparameter   1.05623e+01    2.07058e-01    1.27229e-04   6.80348e-04
   3   Cparameter   5.43143e+02    2.96073e+01    1.74231e-02  -5.65817e-06
   4   Dparameter   2.48481e+00    1.05940e-02    2.51951e-06  -1.45925e-02
   5   Eparameter   1.25761e+00    3.91177e-02    2.02622e-05   3.06499e-03
   6   Fparameter   7.60371e+02    1.68194e+01   -1.42273e-02   6.17900e-06
   7   Gparameter   1.38737e+01    1.11765e-02    1.99857e-06   6.95162e-03
   8   Hparameter   5.06301e-01    1.09724e-02    6.25057e-06  -2.69276e-04
   9   Iparameter   1.86216e+03    3.72654e+01    1.20056e-02  -1.38324e-06
  10   Lparameter   1.24976e+01    3.94562e-03   -1.55961e-07  -1.07880e-02
  11   Mparameter   2.08392e-01    3.56730e-03    1.70195e-06   3.93593e-02
Degrees of freedom: 89 Chi-Square: 113.926 Prob: 0.0386369
The reduced chi-square is: 1.28007
Info in <TCanvas::Print>: pdf file secondstep.pdf has been created
root [1]
```

Fig. 5.23 Fitting macro execution by Root interpreter

The code, interpreted by ROOT, shown in Listing 5.18 first minimizes the parameters of each function within their respective fitting ranges. After these preliminary fits, a new function is defined as the sum of all the component functions. This combined function is created after the individual fits to inherit the optimized parameter values from the components. These initial values are already close to the optimal solution and serve as an excellent starting point for the global fit using the sum function.

During execution, ROOT prints the fitted parameter values and their associated uncertainties directly to the terminal. At the end of the process, thanks to the implementation in the macro, it also prints the reduced chi-square value of the global fit, as shown in Fig. 5.23.

While in Fig. 5.24, the result of the partials (colored lines) and global fit(dashed line).

Fig. 5.24 Data solid line, first peak shape fitted with a Breit-wigner (blue line), second peak with a Gaussian (green line), last peak with a Breit-wigner (yellow line), exponential background (red line), global fit dashed line

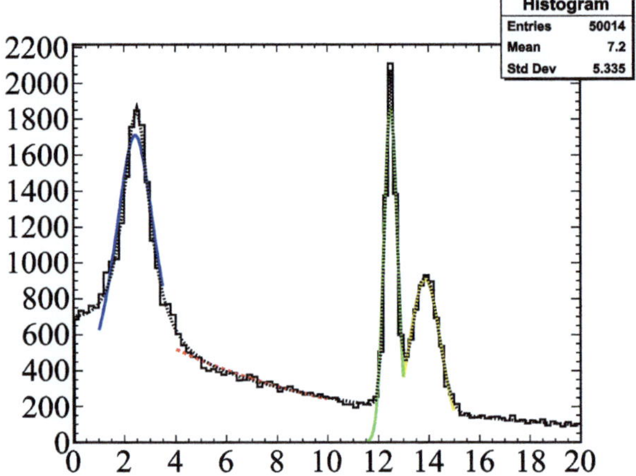

The last step is to recover the background function's parameters from the global fit, give these parameters back to the background's TF1 class, and create a histogram representing the background, as shown in Code 5.19. Finally, it is possible to proceed to background subtraction using the TH1D class's Add method.

```cpp
{
    // To be added to the previous code!

    TCanvas *window2 = new TCanvas("window2", "window2", 600,
    500);
    // Plot the first copy, free from the attributions added
    by the fit
    copy1->Draw();
    double parameters[11];
    //Getting in a vector the parameters of the sum  function
    after the global fit
    sum_function->GetParameters(parameters);
    // Transfer the global fit parameters to the component
    functions
    exponential->SetParameters(&parameters[0]);
    breit_wigner->SetParameters(&parameters[2]);
    breit_wigner2->SetParameters(&parameters[5]);
    gaussian->SetParameters(&parameters[8]);
    // Modify the range of definitions for the component
    functions
    exponential->SetRange(0, 20);
    breit_wigner->SetRange(0, 20);
    breit_wigner2->SetRange(0, 20);
    gaussian->SetRange(0, 20);

    exponential->Draw("same");
    window2->SaveAs("databackgd.pdf");
    TCanvas *window3 = new TCanvas("window3", "window3", 600,
    500);
```

Fig. 5.25 Histogram after
the background subtraction

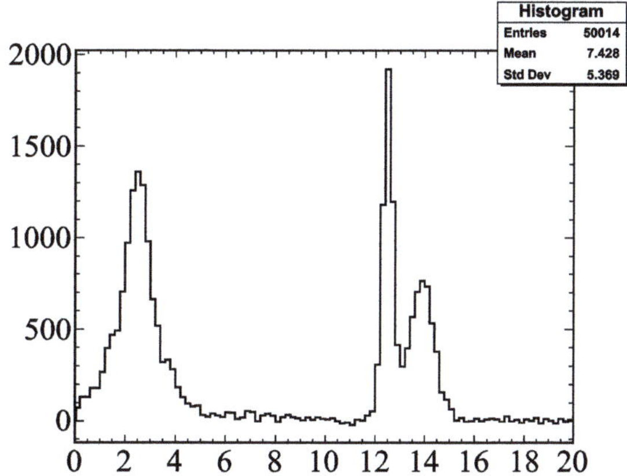


```
24    // Perform the background subtraction!!!!!
25    copy2->Add(exponential, -1);
26    copy2->Draw();
27    //copy2->Draw("same");
28    window3->SaveAs("signals.pdf");
29    TCanvas *window4 = new TCanvas("window4", "window4", 600,
      500);
30    // Plot the components
31 //    exponential->Draw("same");
32    breit_wigner2->SetLineColor(2);
33    gaussian->Draw();
34    breit_wigner->Draw("same");
35    breit_wigner2->Draw("same");
36     window4->SaveAs("models.pdf");
37 }
```

Code 5.19 Additional code to be included in Code 5.18 to pass the parameters estimated by the global fit to the component functions

By subtracting the exponential background from the three peak signals, we obtain the result we have reported in Fig. 5.25.

Looking at Fig. 5.25, we see that the first peak is entirely isolated. This allows us to directly integrate the histogram in the interval from 0 to 6 to obtain a reliable estimate of the counting rate for that signal. However, the situation is different for the second and third peaks, which partially overlap—each acting as background noise for the other.

How can we address this issue? We've solved it by performing a global fit considering all individual contributions.

To estimate the counts for the second and third peaks, we can proceed in two ways: integrate their respective model functions directly, using the parameters obtained from the global fit (represented by the green and red curves in Fig. 5.26); or apply a background subtraction method, similar to the one used earlier, to remove the exponential background.

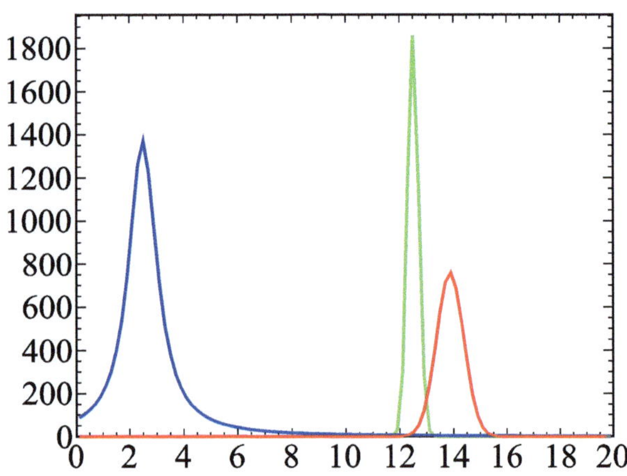

Fig. 5.26 Functions of all signals

We encourage the reader to try both approaches to estimate the counting rates for the second and third peaks and to compare the results obtained with each method.

5.3.6   Challenge Fit

Let us now suppose that we have carried out an experiment examining the relationship between the angle at which a photodetector registers a signal and the corresponding signal measurement in millivolts (mV). The relevant data are provided in the file datuzzi.dat and summarized in Table 5.2.

Let's assume that the law defined by function Eq. 4.1 governs the relationship between the two measured quantities, and that we can use it to perform a fit. The

Table 5.2   Table of data stored in the file "datuzzi.dat"

θ_i (rad.)	R_p (mV)
0.441868	79.10
0.616312	53.20
0.790757	18.70
0.860534	9.16
0.965201	2.00
1.03498	7.60
1.06987	13.70
1.13965	43.45
1.20942	101.50
1.31409	249.20

following script 5.20 includes all options set to produce a refined figure. Yet, in practice, the only instructions essential for executing the fit are marked in the source code with three asterisks, from rows 20 to 25. 5.20.

```
1   //code with fit on data without assigning error
2   // Script fitnoerr
3   {
4     int mymarkerstyle=20;
5     float mymarkersize=1.;
6     float mytextsize=0.065;
7     int mytextfont=132;
8     TCanvas *Canvas_1 = new TCanvas("Canvas_1", "Canvas_1"
        ,600,500);
9     Canvas_1->SetFillColor(0);
10    Canvas_1->SetBorderMode(0);
11    Canvas_1->SetBorderSize(2);
12    Canvas_1->SetLeftMargin(0.15);
13    Canvas_1->SetRightMargin(0.07);
14    Canvas_1->SetTopMargin(0.03);
15    Canvas_1->SetBottomMargin(0.16);
16  //  Canvas_1->SetGridx();
17  //  Canvas_1->SetGridy();
18    Canvas_1->SetTickx(1);
19    gStyle->SetOptStat(1111);
20    TF1 *ff=new TF1("ff","[0]*pow((sqrt(1-pow(sin(x)/[1],2))
21                          -[1]*cos(x))/  (sqrt(1-pow(sin(x)
      /[1],2))
22                          +[1]*cos(x)),2)"
      ,25*3.14/180,75*3.14/180);//***
23    //initialization parameters
24    ff->SetParameter(0,2320);//***
25    ff->SetParameter(1,1.6);//***
26    TGraph *pippo= new TGraph("datuzzi.dat");//**
27    pippo->SetTitle("");
28    pippo->GetXaxis()->SetTitleSize(mytextsize);
29    pippo->GetYaxis()->SetTitleSize(mytextsize);
30    pippo->GetXaxis()->SetLabelSize(mytextsize);
31    pippo->GetYaxis()->SetLabelSize(mytextsize);
32    pippo->GetXaxis()->SetTitleFont(mytextfont);
33    pippo->GetYaxis()->SetTitleFont(mytextfont);
34    pippo->GetYaxis()->CenterTitle(1);
35    pippo->GetXaxis()->CenterTitle(1);
36    pippo->GetYaxis()->SetTitleOffset(1.1);
37    pippo->GetXaxis()->SetTitleOffset(1.10);
38    pippo->GetYaxis()->SetTitle("R_{p} (mV)");
39    pippo->GetXaxis()->SetTitle("#theta_{i} (rad.)");
40    pippo->SetMarkerStyle(20);
41    pippo->Draw("ap");
42    pippo->Fit("ff");//***
43    float mean,sigma,mean2,sigma2,chi,ndf;
44    TLegend *leg = new TLegend(0.2,0.7,0.55,0.85,NULL,"brNDC");
45    leg->SetBorderSize(0);
46    leg->SetLineColor(1);
47    leg->SetLineStyle(1);
```

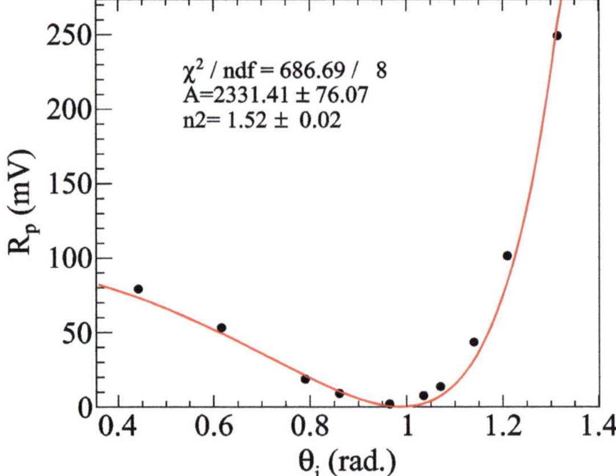

Fig. 5.27 Fit of data

```
48    leg->SetLineWidth(1);
49    leg->SetFillColor(0);
50    leg->SetTextSize(0.05);
51    leg->SetTextFont(132);
52    leg->SetFillStyle(1001);
53    mean=ff->GetParameter(0);
54    sigma=ff->GetParError(0);
55    mean2=ff->GetParameter(1);
56    sigma2=ff->GetParError(1);
57    chi=ff->GetChisquare();
58    ndf=ff->GetNDF();
59    leg->AddEntry(pippo,Form("#chi^{2} / ndf = %6.2f / %3.0f",
         chi,ndf),"");
60    leg->AddEntry(pippo,Form("A=%5.2f #pm %5.2f",mean,sigma),"")
         ;
61    leg->AddEntry(pippo,Form("n2=%5.2f #pm %5.2f",mean2,sigma2),
         "");
62    leg->Draw();
63    Canvas_1->SaveAs("fit.pdf");
64  }
```

Code 5.20 Fit of data reported in Table 5.2

The fit shown in Fig. 5.27 appears to be good—the curve is very close to the experimental points, and the parameters are reasonable for the physics governing this phenomenon. However, the χ^2 variable is large, which generally means the fit is not good. How should we interpret this result? Our assessment of the fit's quality is accurate, and the corresponding χ^2 value was calculated correctly. The apparent discrepancy arises because we did not link the experimental values with their respective errors. In Table 5.3, we display the data under discussion, including the experimental errors related to the angle and those associated with the photodetector's intensity.

Table 5.3 Table of data including measurement errors, stored in the file datuzzierr.dat

θ_i (rad.)	R_p (mV)	err. θ_i (rad.)	err. R_p (mV)
0.441868	79.10	0.0349067	5
0.616312	53.20	0.0349067	5
0.790757	18.70	0.0349067	5
0.860534	9.16	0.0349067	5
0.965201	2.00	0.0349067	5
1.03498	7.60	0.0349067	5
1.06987	13.70	0.0349067	5
1.13965	43.45	0.0349067	5
1.20942	101.50	0.0349067	5
1.31409	249.20	0.0349067	5

Let's redo the fit and modify the previously used script with a minimal change in the code: we switch from the **TGraph** class to the **TGraphErrors** class, which allows us to account for errors on both the x and y variables.

```
// script fitsierr.C
// all as in the previous macro
// only the class that handles
// loading and representing the data changes
// switch from TGraph to TGraphErrors
.....

.....
TGraphErrors *pippo= new TGraphErrors("datuzzierr.dat");
.....
Canvas_1->SaveAs("fiterr.pdf");
```

The result of this second fit is slightly different from the previous one, but is in perfect agreement within the limits of the errors on the parameters we wanted to determine. Meanwhile, the χ^2 is significantly smaller and consistent with our perception of the fit's quality. In this second fit, the minimizing algorithm considered the experimental error, whereas in the first fit, not including this information, it normalized the deviations, factors in the calculation of the statistical variable χ^2, by only the error made in the fit itself. Let's explain this issue in more detail. The fit (the red curve in the figures of this section) has an associated error; it is not error-free. Notice that the parameters A and n_2 have uncertainties in the figure. If we propagate the errors related to these parameters into the formula described by the function (4.1), we obtain the error related to the fit (Fig. 5.28).

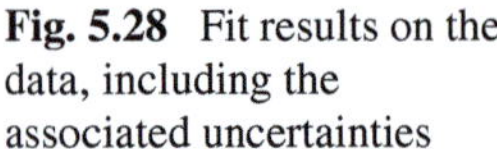

Fig. 5.28 Fit results on the data, including the associated uncertainties

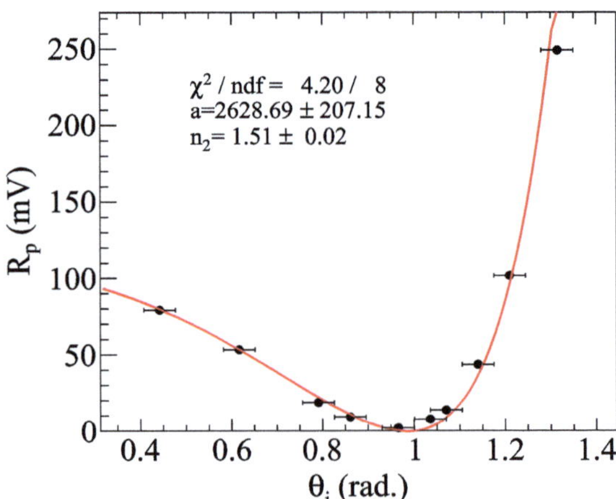

5.4   Data Elaboration and Visualization with Gnuplot

Gnuplot, dating back to 1986, is a powerful graphing utility largely used for data visualization and analysis. It is known for its versatility and ability to generate high-quality plots in various formats. Gnuplot is an essential tool for scientists, engineers, and anyone needing a professional graphing environment and a robust data analysis tool. Gnuplot's core development is supported by a team of volunteers who contribute code, fix bugs, add new features, and ensure that gnuplot remains up-to-date and functional across various platforms. Furthermore there is a large and active user community that provides support through forums, mailing lists and discussion groups.

Gnuplot is compatible with several OS, including Windows, macOS, and Linux. The instructions for installing the latest versions can be found at gnuplot homepage or gnuplot source forge page. It can produce 2D and 3D plots, supporting lines, points, bars, surfaces, and contours. Plots can be extensively customized with different styles, colors, labels, and annotations and it is possible to create animated plots in GIFs.

Gnuplot is essentially a command-line utility, although some GUI interfaces have also been developed. Several programs and programming languages support interfacing with Gnuplot. Being text-driven gnuplot allows for scripting to automate complex plotting tasks or repetitive data analysis flows. Besides plotting ordinary mathematical functions, gnuplot can read data from various sources, including text files, binary files, and directly from SQL databases. It also allows for statistical analysis. Last but not least gnuplot includes fitting capabilities and has support for loops and conditional statements. By leveraging these features, users can create sophisticated scripts that automate plotting, handle multiple datasets, and generate dynamic visualizations based on specific conditions, making it suitable for scientific and engineering applications. As we will show later, these features allow for the solution of simple ordinary differential equations using the Euler method. Even though this richness of functionalities gnuplot has a very smooth learning curve that allows one to master the basics rapidly, then more complex commands can be learned by doing, since everything is explained in the included help system.

Considering its ease of use, gnuplot can also be used as a substitute for spreadsheet-like applications in secondary school to analyze laboratory experiment results or visualize mathematical functions.

5.4.1 Getting Started

After installing the software from the official website or by your system's package manager, the user launches the program by typing `gnuplot` in their terminal or by double-clicking on the gnuplot icon, created during installation. Then a gnuplot shell is opened, ready to accept commands at the prompts: `gnuplot>`. Although only `cd . . . .` and `pwd` shell commands are explicitly included among gnuplot command, it is possible to use any Linux shell command in several ways.

- `! command string`, as in :

```
gnuplot> !ls -l
total 7768
-rw-r--r--  1 ulderico   staff      16661 May 27 16:57 DataSet.odg
-rw-r--r--@ 1 ulderico   staff      17808 May 27 16:57 DataSet.pdf
-rw-r--r--  1 ulderico   staff      34283 Sep 13 17:46 Figure 0.pdf
-rw-r--r--@ 1 ulderico   staff      34859 Sep 13 17:47 Gnu_atan.pdf
-rw-r--r--@ 1 ulderico   staff      40414 May 23 22:27 Gnuplot1.png
-rw-r--r--@ 1 ulderico   staff       9205 May 26 19:59 Mydata.png
-rw-r--r--@ 1 ulderico   staff        826 Aug 28 12:35 cascade.gpl
-rw-r--r--@ 1 ulderico   staff       9402 Aug 28 12:38 cascade.png
-rw-r--r--@ 1 ulderico   staff      20306 Aug 25 11:07 cos2.png
...........................................................................
...........................................................................
```

Code 5.21 Example of gnuplot shell executing a shell command

Or it can be done as well as with `system "ls"`.
This second form can be used to assign values to variables from keyboard input by typing `mychoice = system("read mychoice; echo $mychoice")`.

Gnuplot's shell can also be used as a calculator using the `print` command:

```
gnuplot> print 3+2
5
gnuplot> print exp(42)
1.7392749415205e+18
gnuplot> print 3/2
1
gnuplot> print 3./2
1.5
```

Code 5.22 Example of gnuplot shell used for mathematical operations

Note that gnuplot uses both 'real' and 'integer' and 'complex' arithmetic, like FORTRAN and C. Integers are entered as "1", "–10", etc.; reals as "1.0", "–10.0", "1e1", 3.5e-1, etc. This will affect the results from division, as can be seen in the last two examples.

Complex constants are declared as $\{< real >, < imag >\}$, where $< real >$ and $< imag >$ must be numerical constants. The real and imaginary components of a complex value are obtained by `real(x)` and `imag(x)`, while the modulus is given by `abs(x)`. The imaginary units ι can be obtained written as $\{0.0, 1.0\}$

```
a={1,2}
gnuplot> print real(a)
1.0
gnuplot> print imag(a)
2.0
gnuplot> print sqrt(-1)
{0.0, 1.0}
```

Code 5.23 Definition of a complex number in gnuplot

Since gnuplot can plot only real values, to plot a complex-valued function f(x) one has to choose between plotting real(f(x)), imag(f(x)) or abs(f(x)).

Gnuplot also supports arrays as indexed lists of user variables. Array elements are not restricted to a single type of variable. All elements in an array are initially undefined, and the array size cannot be changed after creation. The number of elements in an array A is given by the expression |A|. Arrays are to be explicitly declared by the `array` command:

```
gnuplot> array A[6]
gnuplot> A[1] = 11
gnuplot> A[2] = 2.0
gnuplot> A[3] = {3.0, 1.0}
gnuplot> A[4] = "Furs"
gnuplot> A[6] = A[2]**2
gnuplot> print A
[11,2.0,{3.0, 1.0},"Furs", ,4.0]
gnuplot> array B[5] = [11,2.0,A[3],"Furs" ,B[2]**2] ; print B
[11,2.0,{3.0, 1.0},"Furs",4.0]
gnuplot> print |A| , |B|
6 5
```

Code 5.24 Definition and use of array in gnuplot

Note that we have used a semicolon in line 9 to put consecutive commands on the same line.

5.4.2 Plotting Functions and Data

To plot a function or data, just requires to type simple commands at the gnuplot prompt. The 'plot' command allows plotting mathematical functions or data points stored in a user file in plain text format. In mathematical expressions, gnuplot plots the function by dividing the x-axis into several points and evaluating the expression at each point. These evaluation values are called "samples." By default, gnuplot will use 100 samples.

```
plot sin(x), 'cos.dat'
```
Code 5.25 Plotting of a mathematical function and data points from a text file

will produce the graph in Fig. 5.29. The example shows how straightforward it is to plot mathematical functions and user data. 'cos.dat' is a simple text file containing two columns with Θ and $\cos(\Theta)$ tabulated values. Math functions can also be user-defined, and the data file may contain several columns (the first and second columns are used by default for x and y). In the previous command, the comma separates various plot instances. The same result can be obtained in another way by

```
plot sin(x)
replot 'cos.dat'
```
Code 5.26 Example of the use of 'replot' command

In the example, the output is rendered on an X11 terminal, using default values for axis ranges, line style, and other graphical parameters. X11 terminal has an interactive window that allows for zooming and varying the coordinate ranges by simple mouse gesture (typing h on the graph will print, in the terminal, helps for possible actions). This simplifies visual data exploration without redrawing, although the output is quite basic and doesn't look very appealing, especially for presentation purposes. On the other hand, the same graph can be easily made more professional by adding some configurations. This is achieved by setting some parameters about plot attributes, the 'set' command is used in this scope with a very intuitive syntax.

```
set terminal aqua enhanced font "arial,18"
set title 'Trigonometric function' offset screen -0.2 font "
    ,22" textcolor rgb "blue"
set zeroaxis linetype -1
set grid
set key box outside left
# primary X in radians
set xrange   [-4*pi:4*pi]
set xlabel   'Angle (radians)'
set xtics    pi    # major tic every π
set format   x  '%.2f'
# secondary X2 in degrees (linked to X)
set link x2 via (x*180./pi) inverse (x*pi/180.)
set x2label 'Angle (degrees)'
set x2tics  180        #tic step
set format  x2 '%.0f'
set ylabel 'Function values'
plot sin(x) title 'sin(x)',   cos(x) w p pt 7 ps 2 lc rgb 'red'
    title 'cos(x)'
```
Code 5.27 Example of plot customization

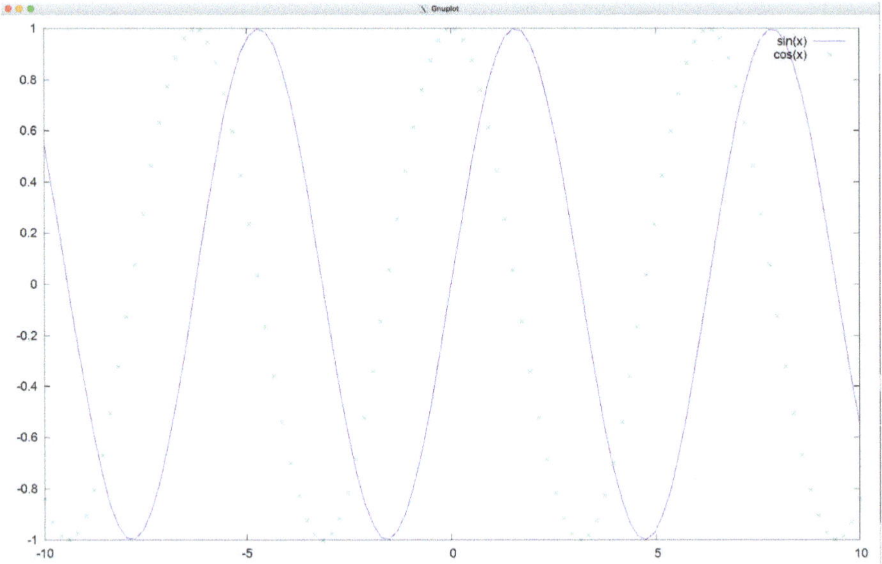

Fig. 5.29 Graphical output of 'plot sin(x)' command

which will produce the graph in Fig. 5.30, where 'aqua' terminal is used. Aqua terminal is less interactive but shows better graphics and saving options (the same is for 'wgt' terminal on Windows installation). The rest of the script illustrates how to set the title (font and position), zero reference axis, grid, axis labels, legend (key) position, and a second x axis (x2) with tic marks in degree, illustrating the possibility of linked axes.

Although some command options may be long to type, most of them can be abbreviated to speed up typing:

```
1  set title 'Trigonometric function' offset screen −0.5 font "arial,22" textcolor rgb "blue"
2  set tit 'Trigonometric function' off sc −0.5 font "arial,22" tc rgb "blue"
3  ......
4  plot sin(x) title 'sin(x)', 'cos.dat' with points pointtype 15 title 'cos(x)'
5  plot sin(x) t 'sin(x)', 'cos.dat' w p pt 15 t 'cos(x)'
```

Code 5.28 Example of commands abbreviations

Gnuplot supports many output formats on several terminals, including graphical files like PNG, PDF, jpg, LaTeX, tkz, and pstricks; type 'set terminal' for a complete list. By setting a terminal type, the plot output will be formatted accordingly. In such cases, it is often necessary to redirect the plot output to a file with the correct extension; otherwise, it will be sent to the gnuplot's window, flooding it with text or binary code according to the used graphical format. For example, to get the plot in Fig. 5.29 as png file, one should input the following commands:

```
1  set term png            # or any other graphic format
2  set out 'myplot.png'     # filename with matching extension
```

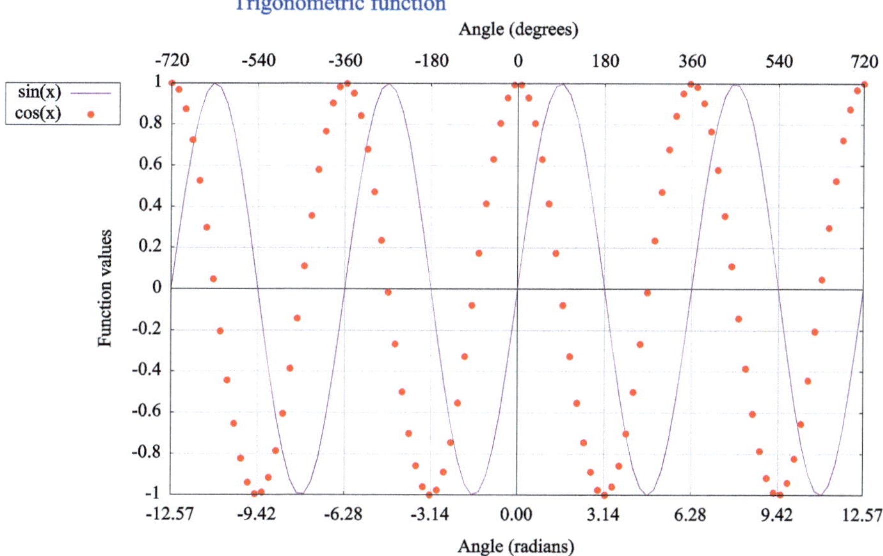

Fig. 5.30 Same plot as in Fig. 5.29, with some customizations

```
3 plot sin(x), 'cos.dat'
4 set out                    # return to standard output
5 set term x11               # return to preferred terminal
```
Code 5.29 Saving a plot as graphic file with the gnuplot shell

It is also possible to send the plotted data or functions in tabular form to a text file, writing both abscissa and ordinates of each plotted line as a column pair. This is achieved by the commands 'set table < outfile >':

```
1 set table 'mycos.txt'    # create or overwrite the output file
2 p [-8:- 6] 'cos.dat'     # [-8:- 6] fix the x range plot limits
3 unset table              # return to standard output
```
Code 5.30 Saving a plot as text file with the gnuplot shell

which will produce the output in Fig. 5.31. If 'outfile' exists, it will be overwritten unless the 'append' keyword is given after < outfile > set table 'mycos.txt' a. In this example, we have limited the x range of the 'plot' command to highlight the flag character that follows the tabulated data. It will be 'i' if the point is in the active range, 'o' if it is out-of-range, or 'u' if it is undefined.

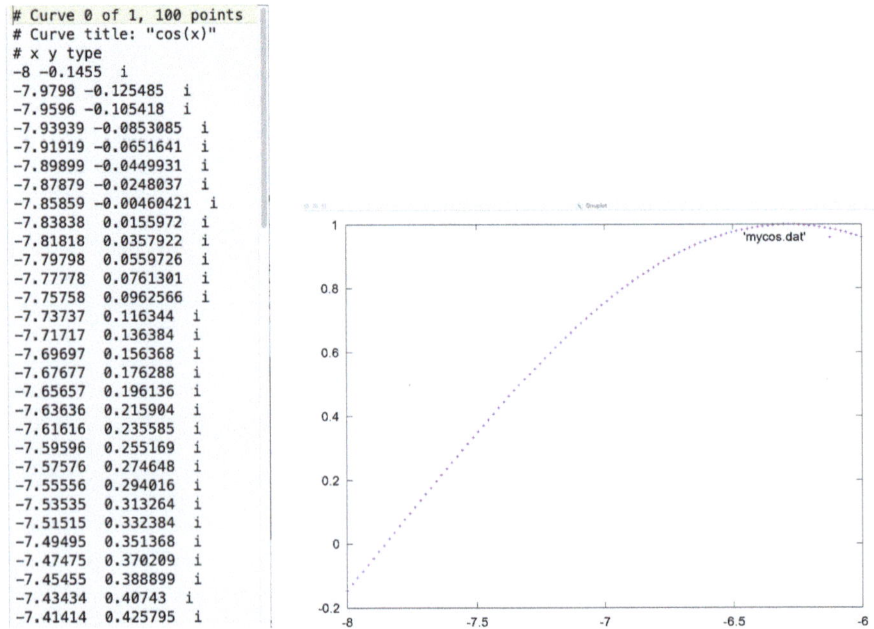

Fig. 5.31 The 'tabled' output file of 'plot' command and the X11 window output with the same command

By using `set out 'filename'` or `set table 'filename'` it is then possible to redirect the 'plot' output but also the 'print' output (default to screen) can be redirected to a file by `set print 'filename'`, with or without 'append' option as for 'set table'. In this case, the format of each output line is that of the corresponding 'print' command.

```
gnuplot> print '1 2 3 4 5 bar  foo'
1 2 3 4 5 bar  foo
print sqrt(2.)
1.4142135623731
gnuplot> set print 'useless.dat' # redirect print to file
gnuplot> print '1 2 3 4 5 bar  foo'
gnuplot> print sqrt(2.)
gnuplot> set print  # go back to standard output (screen)
gnuplot> !cat useless.dat  # execute shell command 'cat'
1 2 3 4 5 bar  foo
1.4142135623731
gnuplot>
```

Code 5.31 Redirecting gnuplot printed output to a file

As can be seen, most of gnuplot's commands are almost self-explanatory, even for non-English speakers, and sometimes they can even be guessed. Just in case the command 'unset < something >' is used for returning < something > to its default state. In any case, a handy `help` command recalls the user about the command syntax and options.

```
gnuplot \textgreater help
'Gnuplot' is a portable command-line driven graphing utility for Linux, OS
    /2,
MS Windows, OSX, VMS, and many other platforms. The source code is
    copyrighted
but freely distributed (i.e., you don't have to pay for it). It was
    originally
created to allow scientists and students to visualize mathematical
    functions
and data interactively, but has grown to support many non-interactive uses
such as web scripting. It is also used as a plotting engine by third-party
applications like Octave. Gnuplot has been supported and under active
development since 1986.
..........
Help topics available:
    2D               3D               Features         arrays
    automated        autoscale        background       batch/interactive
    beeswarm         binary           bugs             call
    canvas           changes          circle           colornames
    colorspec        commands         comments         complete
    complex          constants        coordinates      dashtype
    data-file        datafile         datastrings      demos
    deprecated       ellipse          elliptic         enhanced
    environment      expressions      fenceplots       fillcolor
    filledcurves     fonts            for              gd
..........
    startup          string           style            substitution
    surface          syntax           time/date        unset
    using            variable         while            xticlabels
```

Code 5.32 Shell output of 'help' command

5.4.3 User-Defined Variables and Functions

In addition to standard mathematical functions that are already defined, users can define local variables and functions that can also be used in the plot command. Functions are limited to twelve parameters and include string functions. Generally, any mathematical expression accepted by C, Python, and FORTRAN may be used, and most operators accept integer, real, and complex arguments. Note that exponentiation is done through the ** operator (as in FORTRAN)

As an example (adapted from gnuplot >help user-defined) we have:

```
w = 2   # simple assigment
q = floor(tan(pi/2 - 0.1))
f(x) = sin(w*x)
sinc(x) = sin(pi*x)/(pi*x)
delta(t) = (t == 0)
ramp(t) = (t  > 0) ? t : 0  # if t > 0 then t else 0
min(a,b) = (a <  b) ? a : b
comb(n,k) = n!/(k!*(n-k)!)
```

```
 9  len3d(x,y,z) = sqrt(x**2 +y**2 + z**2)
10  plot f(x) = sin(x*a), a = 0.2, f(x), a = 0.4, f(x)
11  file = "mydata.dat" # string variable
12  file(n) = sprintf("run_
```
Code 5.33 Definition of mathematical functions in gnuplot

Furthermore, the variables pi (3.14159...) and NaN (IEEE "Not a Number") are already defined.

All these definitions are supposed to be on a single line, which can be possibly broken by the line continuation character '\'. Multi-line functions can be defined using the function block mechanism that allows for a named gnuplot code block to be called as function.

5.4.3.1 Parametric Plot

Besides plotting ordinary functions with an explicit relationship between the x and y values, it is possible to plot parametric functions where the x and y values depend on a third variable, called a parameter t. This is achieved with the set parametric command as in the example below:

```
1  set samples 500  # increase sampled points
2  set parametric
3  unset key
4  set trange  [-2*pi:2*pi] # define range for t
5  plot sin(2*t), cos(3*t)  # plot Lissajous figure
```
Code 5.34 Definition of parametric functions: Lissajous figure, in gnuplot

That will produce the plot on the left side of the Fig. 5.32. In the same way, we can draw a spiral and a circle by typing:

```
 1  reset
 2  set size  square  # control the aspect ratio of the plot
        window
 3  set xzeroaxis # draw y=0 axis
 4  set yzeroaxis # draw x=0 axis
 5  set key out   # set legend key outside the plot window
 6  set parametric
 7  set xrange[-5*pi:5*pi]
 8  set yrange[-5*pi:5*pi]
 9  plot[0:5*pi] t*cos(t), t*sin(t)  w lp # spiral
10  rep  5*pi*cos(t), 5*pi*sin(t) #circle of radius 2$\pi$
```
Code 5.35 Definition of parametric functions: spiral and circle, in gnuplot

whose output is shown on the right side of Fig. 5.32. In this case, we have also adjusted the aspect ratio of the plot windows and moved the legend outside on the right.

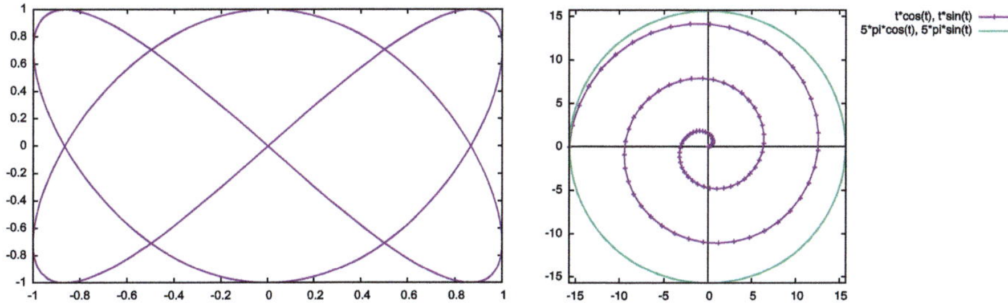

Fig. 5.32 Parametric plots: Lissajous figure on the left and spiral and circle on the right

5.4.4 2D Plotting from User Data

Plotting data sequences, like measurement results ordered in columns, works basically as the plotting of functions. But in this case, we need a file with our data and commands to specify which column has to be the 'x' and which column has to be the 'y' (or error, data label ...). This is done by the 'using n:m' specification to indicate using the nth column as 'x' and the mth column as 'y'. When not specified, the first and second columns are chosen.

Consider, as an example, a data file of the following type:

```
 1  # Time           Current            Error              Background
 2  0.20202          0.59645            0.049104           0.22843
 3  0.40404          0.98018            0.056174           0.16592
 4  0.606061         0.90805            0.048060           0.20787
 5  0.808081         0.57372            0.043970           0.19996
 6  1.0101           0.04680            0.062735           0.18387
 7  1.21212         -0.40784            0.059054           0.11693
 8  1.41414         -0.71141            0.025983           0.14476
 9  1.61616         -0.75434            0.031580           0.07771
10  1.81818         -0.49922            0.055977           0.13190
11  2.0202           0.02861            0.041060           0.03491
12  2.22222          0.33590            . . . . . . .       . . . . .
13  . . . . . . .    . . . . . . . .
```

Code 5.36 Example of data file in multi coulumns format

where '#' is a comment character and entries are separated by white space, that are the default option. These defaults can be changed to the user's preferences by

```
1  set datafile separator {whitespace | tab | comma | "<chars>"}
2  set datafile commentschars {"<string>"}
```

Code 5.37 Selecting file separator and comments characters in text file

To plot current versus time, we use:

```
1  p 'mydata.dat' w lp          # short for  using 1:2 with
      linepoint
```

Code 5.38 Example of plot from multi coulumns data file

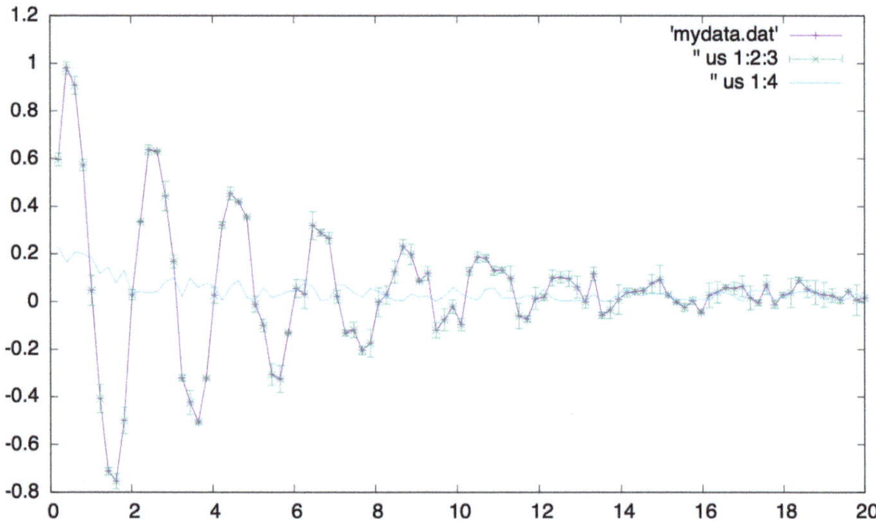

Fig. 5.33 Plot of an experimental data set with error bars and a background

Then to add the error (col 3), we add a second plot instance, using the option 'with error'

```
p 'mydata.dat' w lp, '' us 1:2:3 with error
```

Code 5.39 Example of plot with errors from multi coulumns data file

where the two apexes mean: keep using the same data file. And finally, to add the background (col 4)

```
p 'mydata.dat' w lp, '' us 1:2:3 w err, '' us 1:4 w l
```

Code 5.40 Ploting several coulumns from data file

This last command will eventually produce:

Gnuplot can also read and plot from multiple datasets, i.e., contiguous data files separated by two blank lines. As an example, we can split the previous 'mydata.dat' file into three subsets, as shown in Fig. 5.34. As can be noticed, datasets are enumerated starting from 0, and in the 'plot' command, it is necessary to specify which dataset has to be plotted. This can be achieved by adding the specification 'index $=<$ integer $>$' to the plot command

```
plot 'mydata-split.dat' index 2 w lp  # with linepoint (w lp)
replot '' i 1            # same file with point (default)
rep '' i 0 w lp,'' i 0 us 1:2:3 w e # linepoint and error (w e)
```

Code 5.41 Example of script to plot from blocks data file

The three plot instances can be typed in a single line, separated by a comma. In any case, the output will be that shown in Fig. 5.35.

Of course, for many occasional users of gnuplot, it may not be easy to remember all the available point types or line styles and colors, which may also depend on

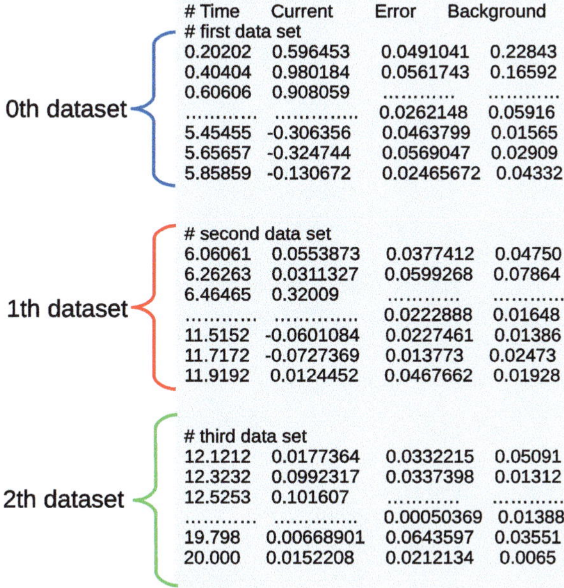

Fig. 5.34   Example of a gnuplot multiple data set

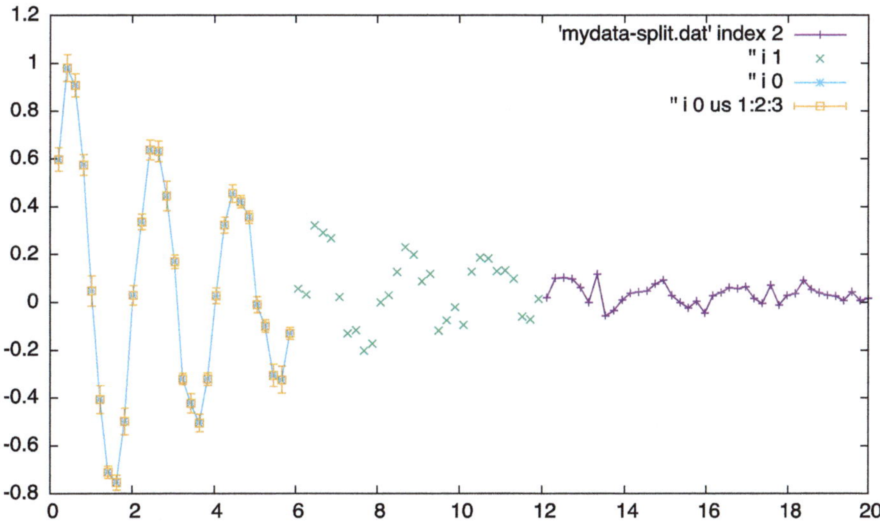

Fig. 5.35   Example of plots from the blocks data file

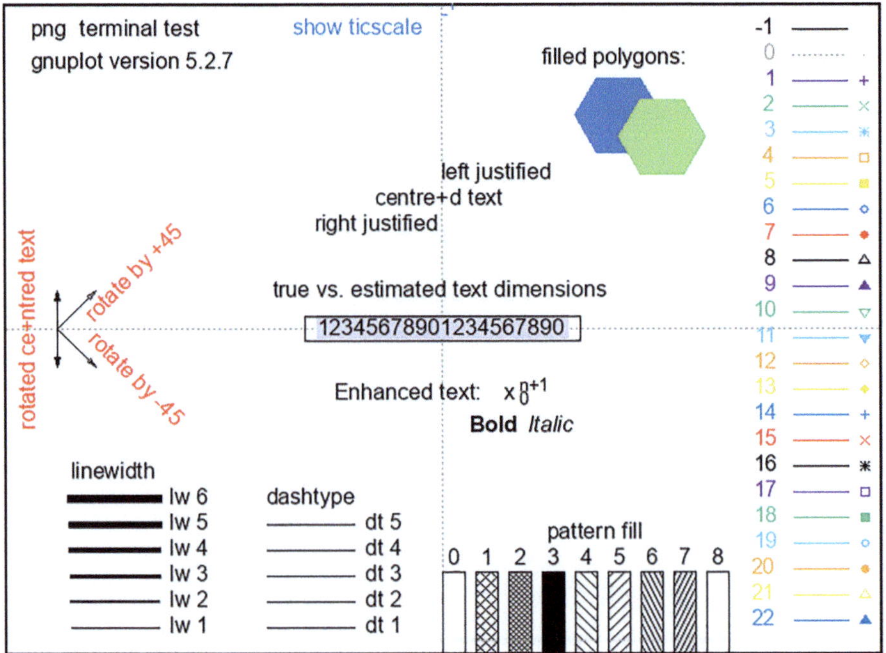

Fig. 5.36 Plot output produced by the 'test' command

the type of terminal used. In this case, the 'test' command will help by plotting a miscellany of the available options, as in Fig. 5.36.

5.4.5   Operation Among Columns

One of Gnuplot's most useful functions for data analysis is the possibility of plotting a data file while making mathematical and logical operations among columns on the fly. This is achieved by adding the $ sign as a prefix to the column number used in a plot command. In this case, the read values are treated as numbers and can be combined in a mathematical expression before being plotted.

The following command

```
p 'anydatafile'  us ($1-3):(sqrt($2)+ 1/$3)
```

Code 5.42 Plot of mathematical operations among coulumns

will produce a plot where the abscissa is the first column, in the file 'anydatafile', and it is shifted to the left by three units, while the ordinate is the square root of the second column plus the inverse of the third column.

For example, we can refer to the previously used cosine tabulated values written in the 'cos.dat' file.

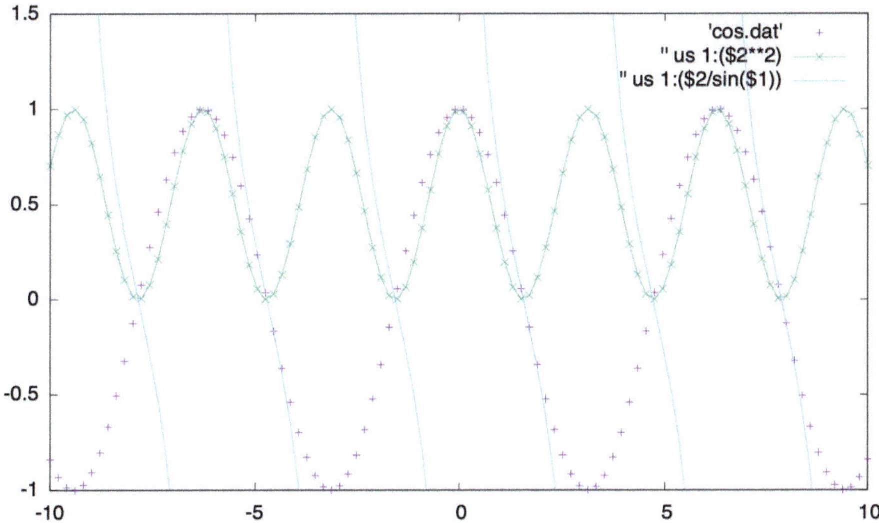

Fig. 5.37 Example of plots obtained by mathematical operation on column data files

By typing

```
set yrange [-1:1.5]  # to leave room for the legend (key)
p 'cos.dat', '' us 1:($2**2) w lp, '' us 1:($2/sin($1)) w
  l
gnuplot>
```

Code 5.43 Plot of mathematical function using coulumns values as argument

We obtain the plot in Fig. 5.37, where, besides the tabulated cosine data points, we plot the square of the data (with line points) and the tangent function by using the sine of the first column and the tabulated cosine values in the second column (with line).

This allows one to quickly test any complex transformation on data, leaving the original data unchanged and saving the transformed data as needed. Moreover, the different operations applied to a data file can be saved in scripts that can be later used to reproduce the same procedure on similar data sets.

5.4.6 Multiplot

Different plots on the same graph with various sizes and positions are often necessary. This is achieved by the 'set multiplot' command, which opens an environment where the origin (bottom-left position) and the size (width and height) for each single plot can be defined prior to plotting. This environment is terminated by 'unset multiplot'.

Both dimensions are in 'screen' coordinates in which the bottom-left corner has
coordinate 0,0 and the top-right corner has coordinate 1, 1.

```
set origin 0.25,0.25
set size 0.5,0.5
```

Code 5.44 Setting origin and size for multiplots

will produce a centered plot of a quarter in size.

For an optimal choice about positions and size of different plots, it is better to
write a script that can be quickly updated and executed to evaluate the final result.

As an example, we can prepare a script to plot some of the graphs that we have
shown previously collected in a single figure.

```
reset                  # unset previous settings
set multi              # short form of multiplot
# BOTTOM
set size 1.0,0.5    # full orizontal width and half vertical
     heigh
set origin 0.0,0.0 # reduntant (default values)
p 'mydata.dat' w lp, '' us 1:2:3 w err, '' us 1:4 w l
# TOP LEFT
set origin 0.0,0.5
set size 0.5,0.5   # half orizontal width, half vertical heigh
set yrange [-1:1.5] # to have room for the legend (key)
p 'cos.dat', '' us 1:($2**2) w lp, '' us 1:($2/sin($1)) w l
unset yrange
# TOP RIGHT
set origin 0.5,0.5
p [-8:- 6]  'cos.dat' pt 5 #
unset multi
reset
```

Code 5.45 Plotting data and functions on four three panels

Let's save this script as 'multi.gpl' in the same directory as the used data file.
Then we can load it within gnuplot terminal by:

```
gnuplot> load 'multi.gpl'
```

Code 5.46 Loading and executing a script with gnuplot commands

We eventually obtain the picture in Fig. 5.38, which can be easily modified by
updating the script and reloading it from the gnuplot terminal, using the up arrow
key to recover past commands.

Multiplot mode can also be used to overlap graphs to produce, as example, a
cascade plot.

```
reset
set sample 200
unset key
lor(x,w)=w/(w**2 + x**2) # lorentzian function definition
set multi
set yrange [0:2]
w= 0.5
dx= 0.
dy=0.
set border 1+2 # plot only bottom x1 an y1 axes
```

```
11  p lor(x,w)
12  w=w*1.4            # increase the width a bit
13  dx= dx+0.05
14  dy=dy+0.05
15  set origin dx,dy # shift the origin a bit
16  p lor(x,w) lc 2
17  w=w*1.4            # increase the width a bit
18  dx= dx+0.05
19  dy=dy+0.05
20  set origin dx,dy # shift the origin a bit
21  p lor(x,w) lc 3
22  w=w*1.4            # increase the width a bit
23  dx= dx+0.05
24  dy=dy+0.05
25  set origin dx,dy # shift the origin a bit
26  p lor(x,w) lc 4
27  w=w*1.4            # increase the width a bit
28  dx= dx+0.05
29  dy=dy+0.05
30  set origin dx,dy # shift the origin a bit
31  p lor(x,w) lc 6
32  w=w*1.4            # increase the width a bit
33  dx= dx+0.05
34  dy=dy+0.05
35  set origin dx,dy # shift the origin a bit
36  p lor(x,w)  lc 7
37  unset multi
38  reset
```

Code 5.47 Example of script for a cascade plot, using multiplot

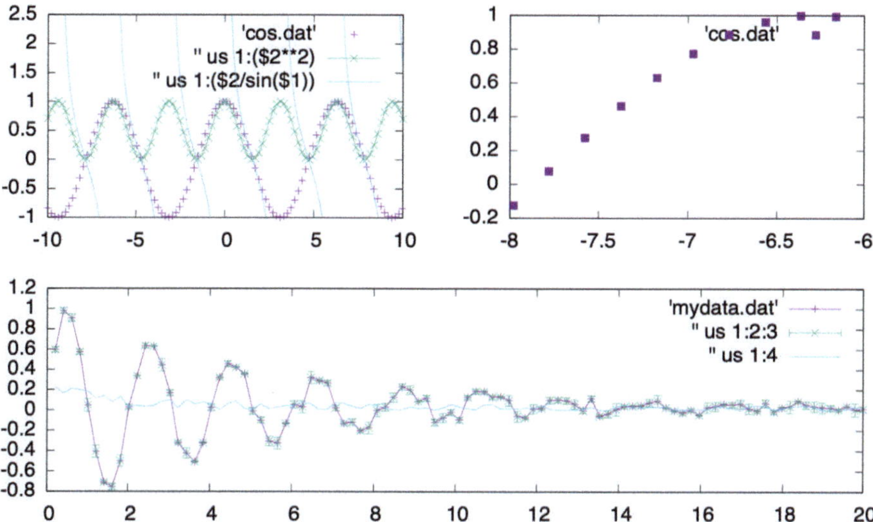

Fig. 5.38 Some of the previous graphs plotted together using 'multiplot' environment

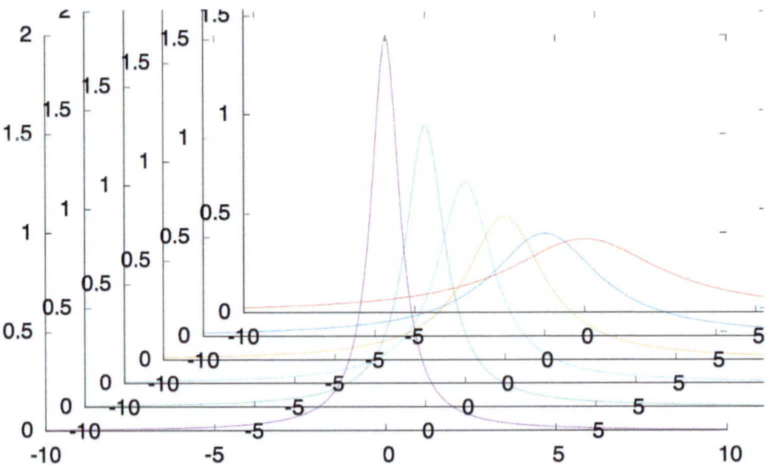

Fig. 5.39 Example of cascade plot, using multiplot environment

The output is shown in Fig. 5.39, where several lorentzian curves with increasing line width are plotted with an increasing origin shifts. We will later see how to iterate actions within gnuplot using do loops.

The multiplot environment can also be used to produce broken-axis plots with two disjoint ranges on one axis by aligning or stacking two plots side by side.

5.4.7 Fitting

In Gnuplot, the 'fit' command just fits a mathematical expressions to a dataset via some adjustable parameters. It uses the nonlinear least-squares Marquardt-Levenberg algorithm. The command syntax is quite obvious:

```
fit MathExpression(x) 'datafilename' us 1:2 via a,b,c .....
```

Code 5.48 Sintax of the 'fit' command

where MathExpression(x) is the expression to be tested and a, b, c ..., are any of its adjustable parameters. Similarly to the 'plot' command, the using (us m:n) option allows any combination of columns in the dataset to be fitted. At the same time, the 'via' qualifiers select the free parameters or a file name containing the starting values. Of course, all the parameters implied in the MathExpression(x) definition have to be assigned before starting the fit.

As an example, we may consider the data in Fig. 5.33 that resembles a damped oscillating signal over a noisy background. In this case an appropriate expression would be something like $fitfun(x) = A \sin(2\pi x/T + b) \exp(-x/\tau)$, where A is an amplitude, *T* the oscillation period, *b* a phase shift, and *tau* a decay time, are the adjustable parameters. Of course, a set of reasonable guess values has to be assigned to these parameters to be used as a starting point for the fitting procedure. Those values are often easily obtained by inspecting the experimental data.

As mentioned before, a reasonable starting point is very important in any least squares fitting procedure, and it is always useful to plot the experimental points together with the hypothesized fit function to check their closeness.

It will be sufficient to type the following on a gnuplot terminal to do this.

```
fitfun(x)= A*sin(x*2*pi/T+b)*exp(-x/tau)
A=1.
b=.00001
T=2.
tau=8.
p 'mydata.dat' us 1:($2-$4), fitfun(x) # comparing guess
    function with experiment minus background
```

Code 5.49   Example of defining a fit function and testing starting parameters

Then to actually let gnuplot to find the best set of parameters that minimize the squared difference between our function and the measured signal (2th columns) minus the measured background (4th columns), we will just have to type

```
fit fitfun(x) 'mydata.dat'  us 1:($2-$4) via A,b,T, tau #
    execute the fit with all parameters free
..........
p fitfun(x), '' us 1:($2-$4) # plot fit result and
    experimental data
```

Code 5.50   Fit of experimental data file and plot of the results

The final plot is shown in Fig. 5.40 along with the plot of the guessed starting point.

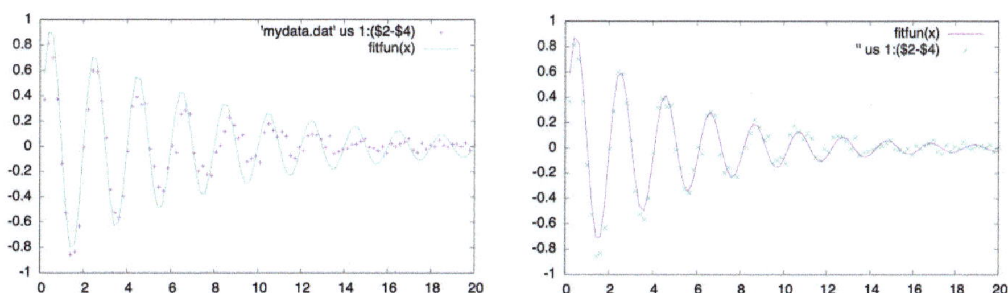

Fig. 5.40   The left side shows a plot of the experimental data (points) and the fitting function (line) with guessed starting parameters. The fitted function is then plotted on the right side with the final set of parameters

At the same time, we will find several information on the gnuplot terminal:

```
iter        chisq        delta/lim  lambda   A              b              T
              tau
   0 1.6352111749e+00    0.00e+00   2.64e+00   1.00e+00    1.00e-05
     2.000000e+00    8.000000e+00
   1 6.7363051924e-01   -1.43e+05   2.64e-01   9.011870e-01   9.999997e-06
     2.018395e+00    6.908587e+00
   2 4.6606355338e-01   -4.45e+04   2.64e-02   9.845289e-01   9.999983e-06
     2.032827e+00    4.837404e+00
   3 4.2907295365e-01   -8.62e+03   2.64e-03   9.805141e-01   1.001316e-05
     2.039973e+00    5.357003e+00
     .................
     .................
     .................
   8 4.2490107485e-01    2.61e-11   2.64e+07   9.842117e-01   4.053820e-02
     2.045222e+00    5.332201e+00
   9 4.2490107485e-01    0.00e+00   2.64e+06   9.842117e-01   4.053820e-02
     2.045222e+00    5.332201e+00
iter        chisq        delta/lim  lambda   A              b              T
              tau

After 9 iterations, the fit converged.
final sum of squares of residuals: 0.424901
rel. change during last iteration : 0

degrees of freedom    (FIT_NDF)                        : 95
rms of residuals      (FIT_STDFIT) = sqrt(WSSR/ndf)    : 0.0668778
variance of residuals (reduced chisquare) = WSSR/ndf   : 0.00447264

Final set of parameters:          Asymptotic Standard Error
=======================           ==========================
A               = 0.984212        +/- 0.03739      (3.799)b = 0.0405382
    +/- 0.04092 (100.9)
T               = 2.04522         +/- 0.007007     (0.3426)tau = 5.3322
    +/- 0.2886 (5.412)

correlation matrix of the fit parameters:
              A      b      T      tau
A             1.000
b            -0.113  1.000
T            -0.086  0.738  1.000
tau          -0.716  0.090  0.074  1.000
```

Code 5.51 Terminal output of the 'fit' command

It will show the progress of minimization: iteration steps, the Chi-squared, delta/lim, and lambda (relative increment and damping factor used by the Levenberg–Marquardt algorithm), and eventually, the fitting parameter values. When convergence is reached, the final values are written together with the standard errors and the correlation matrix. The same information is also written to a log file, "fit.log", and the final values can be saved in a file for subsequent use as a parameter file by using 'save fit'.

The fitting function can contain up to 12 independent variables and any number of adjustable parameters. Optionally, error estimates can be input for weighting the data points, while the fit range may be specified in the same way used in the plot command:

```
1  fit [xmin:xmax] [ymin:ymax] fitfun(x) 'filename' us xcol:ycol:yerrcol yerr
       via par1, par2,....
2  fit [xmin:xmax] [ymin:ymax] fitfun(x) 'filename' us xcol:ycol:xerrcol xerr
       via par1, par2,....
3  fit [xmin:xmax] [ymin:ymax] fitfun(x) 'filename' us xcol:ycol:xerrcol:
       yerrcol xyerr via par1, par2,....
```

Code 5.52 Various forms of the 'fit' command with limits and error weight

Once a set of data has been fitted, it is useful to examine the residuals (the difference between exp. data and fitted curve), which are hopefully as small as possible. A simultaneous view of fitted data and fit residuals requires a broken plot. This can be attained in the multiplot environment, by stacking data plot and residuals plot one on top of each other and removing the bottom axis from the top plot and the top axis from the bottom plot:

```
1  reset
2  #BOTTOM PLOT
3  set sam 200 # to sample plotted functions with 200 points
4  set multi
5  set origin 0,0
6  set size 1,0.2
7  set border 1+2+8 # 1= bottom, 2= left, 4=top, 8=right
8  set xtics nomirror # avoid mirroring x tics on x2 axis
9  set ytics  0.1 # fix y tics interval
10 p 'mydata.dat' us 1:(($2 -$4)-fitfun($1)) w i lt 3 lw 3 not #
       residuals plotted with impulses line width 3  and no title

11
12 #TOP PLOT
13 set size 1,0.8
14 set border 14 # =2+4+8 (left+top+right)
15 unset xtics
16 set origin 0,0.2
17 p 'mydata.dat' us 1:($2 -$4) pt 6, fitfun(x), NaN t 'residues'
       , A*exp(-x/tau) # plot data, fitted function , 'nothing'
       but with a title in the key and the exponential part of
       the fitting function
18 unset multi
```

Code 5.53 Plotting the fit results and fit residues

The above script will produce the graph in Fig. 5.41.

Although this is not properly a broken plot since the y-axis ranges overlap, it has all the necessary elements to be done properly. In particular, it is important to have the same spacing on the left axis, which means that any labels and tics format for the y-axis must take the same room in both plots. That is why we have set the same tics interval on both plots. We have also shown a way to add an item in the legend key without plotting anything (NaN), but with a title so that the label 'residuals' appears in the top plot. As a consequence, we have to plot the residuals in the bottom plot with the same line type as the NaN plot, which is the third type since it is the third draw in the plot command. We have also seen how to select the plot borders encoded in a 12-bit integer: the four low bits (1, 2, 4, 8) control the border for 'plot'. Higher-order bits (16, 32, 64, 128 ... 4096) are used for 3D plots and polar plots.

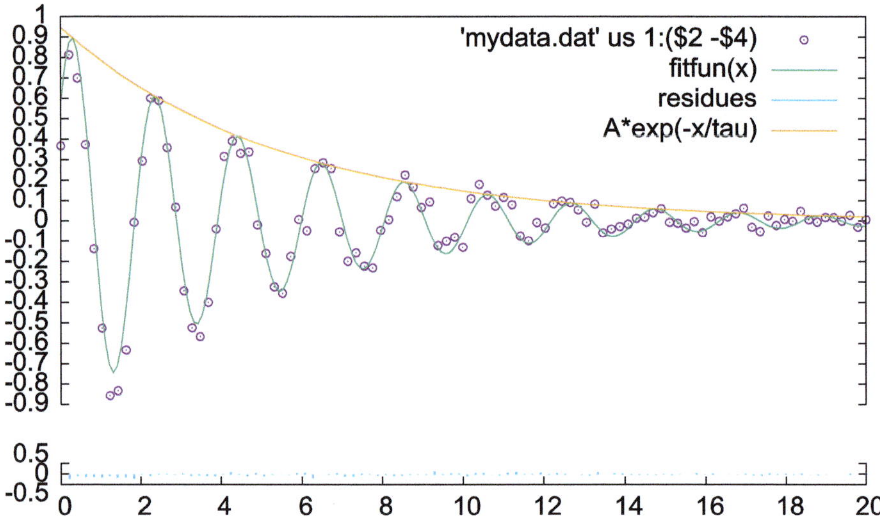

Fig. 5.41 Plot of the experimental data (points), the fitted function and the exponential component (lines). On the bottom side the residues are plotted as vertical bar

There are many more options to the fit command, see 'help fit' within gnuplot for detailed use. In particular, the `set fit prescale` is very useful to scale fit parameters by their initial values; this helps the Marquardt-Levenberg routine converge more quickly and reliably in cases where parameters differ in size by several orders of magnitude. Moreover, it is possible to simultaneously fit the same function to multiple data sets, sharing a common independent variable, with the multi-branch fitting options.

Unfortunately, a feature missing from the Gnuplot fitting routine is the possibility of constraining the range of parameters variation. This is useful when we know (or desire) that certain parameters values are bound within a given range. On the other hand, in a fitting procedure, all the parameters must be internally unlimited for the algorithm to work efficiently. This can be easily overcome by using a variable transformation on our critical parameter using some naturally limited function. The most common choice is the arctangent function, which is limited to be $-\pi/2 < \arctan(x) < \pi/2$, $\forall x \in [-\infty, \infty]$. Considering our previous fit example, we can limit the range of A parameters between A_{min} and A_{max} by rewriting its definition as follows

```
fitfun(x)= A(a)*sin(x*2*pi/T+b)*exp(-x/tau)
# Restrict A in the range  [A_min:A_max]
A(x)  = (A_max-A_min)*(atan(x)+pi/2)/pi + A_min
A_min= 0.5 # minimum value
A_max= 1.5 # maximum value
b=.00001 ; T=2. ; tau=8.
fit fitfun(x) 'mydata.dat'  us 1:($2-$4) via a ,b,T, tau #
    note that now 'a' is the free parameter
```

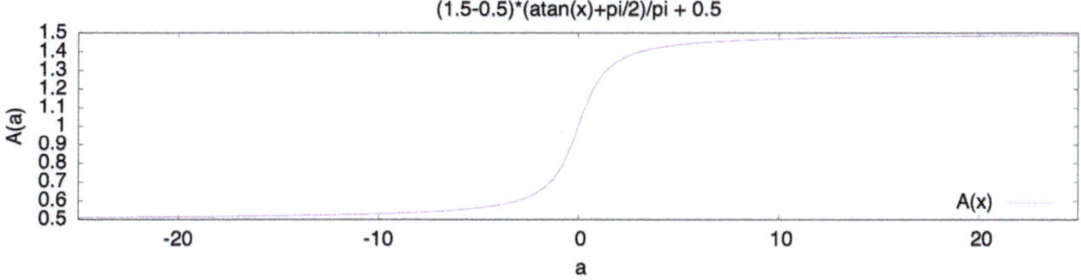

Fig. 5.42 Illustration of the use of 'atan' function to limit a parameter value in a fitting routine

Code 5.54 Example of variable transformations to limit fit parameters

In this case a is running from $-\infty \rightarrow +\infty$ while A(a) is limited as can be seen in Fig. 5.42. Of course the fitting routine will just optimize the a value while the A(a) value is obtained by typing print A(a) at gnuplot prompt.

5.4.8 Condition and Loops

Gnuplot's syntax implements two essential structures: conditions like if..then..else and loops like do for i=start, stop, step, either within the plot command or in the terminal window, as well as in a script. The simplest form of if..then..else is achieved by a ternary operator structure: A?B:C, where A is the requested condition. If A is true, then B is chosen and the rest of the structure is ignored; if A is false, then C is chosen. Such operations can also be concatenated to provide an else if condition like the step function in the example below. This simple construct can be used in defining discontinuous functions:

```
f(x) = sin(x) >= 0.5 ? sin(x) : 0.5   # chopped sinus function
g(x) = abs(x-x0)<W ? (1-abs(x-x0)/W): 0   # triangular function
W=1; x0= 5                    # triangular base width and
    centroid
h(x)= x<=0 ? 0 : x<1 ? 1 : x<2 ? 2 : x<3 ? 3 : 4 # step
    function
set sample 500
p [-5:20][-1:5] f(x), g(x), h(x)
```

Code 5.55 Example of 'conditions' in functions definition

whose plots are shown in Fig. 5.43.

The same construct can also be used within a plot command to select which points are to be drawn according to some choice defined in the using: qualifier:

```
set key bottom left
p 'mycos.txt' us 1:($2>0? $2:NaN) w lp # select only y>0
rep 'mycos.txt' us 1:($1>0? $2/2:NaN) w lp # select only x>0
    and plot y(x)/2

```

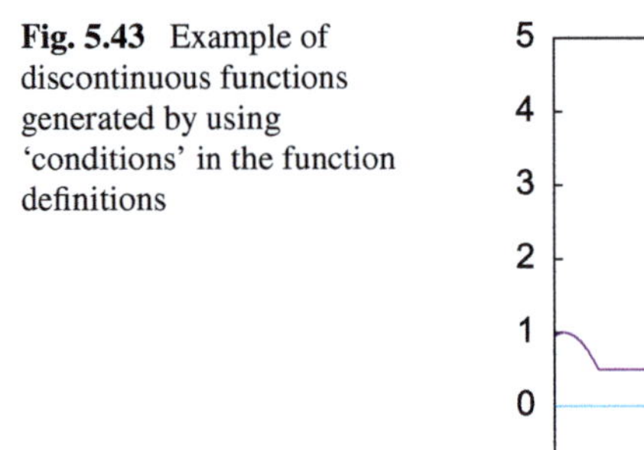

Fig. 5.43 Example of discontinuous functions generated by using 'conditions' in the function definitions

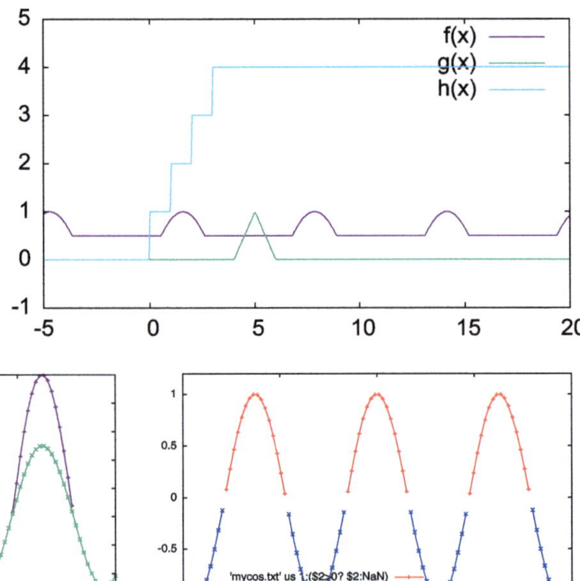

Fig. 5.44 Example of discontinuous plots generated by using 'conditions' in the 'plot' command

```
reset
set yrange [-1:1.2]
p 'mycos.txt' us 1:($2>0? $2:NaN) w lp lc 'red' # select y>0
    and plot them in red
rep 'mycos.txt' us 1:($2<=0? $2:NaN) w lp lc 'blue' # select y
    <0 and plot them in blue
```

Code 5.56 Example of 'conditions' in plot command

whose plots are shown in Fig. 5.44 illustrating the flexibility of the using: qualifier.

The plot and the set commands support an iterative structure in the form for [intvar = start:end:increment] that reiterates the command, reevaluating any expressions that make use of the INT (integer) as a variable. Nested iteration is supported and it also works with string variables in the form: for [stringvar in "A B C D"], as illustrated in the following example:

```
plot for [sample in "A B C"] "Test-".sample.".dat"
```

Code 5.57 Example of the use of 'for' in plot command

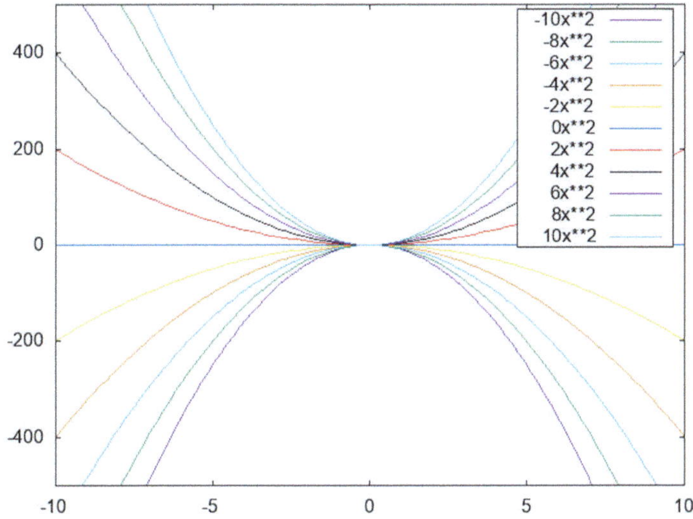

Fig. 5.45 Example of the use of 'for' in plot command

that will plot data from "Test-A.dat", "Test-B.dat" and "Test-C.dat", in sequence.
The example also shows how to concatenate strings using a dot to connect typed text
and a string.

As another example, we can plot a series of parabolas by typing:

```
set key opaque #  legend covering  lines
set key box    #  framed legend
p for [a= -10:10:2] a*x**2 t sprintf("dx**2",a)
```

Code 5.58 Another example of the use of 'for' in plot command

where we have also applied standard C-language format specifiers to write the lines
title (Fig. 5.45).

The iteration of arbitrary command sequences can be obtained using the do command:

```
do for [iteration-spec] {  # curly brackets on do line
      commands
      commands
}                # closing loop
```

Code 5.59 Example of the use of 'do for' iteration

We can use this to simplify the script we wrote for the cascade plot 5.47.

```
reset
set sample 200
unset key
set xtics nomirror ; set ytics nomirror
lor(x,w)=w/(w**2 + x**2) # lorentzian function definition
set multi
set yrange [0:2]
```

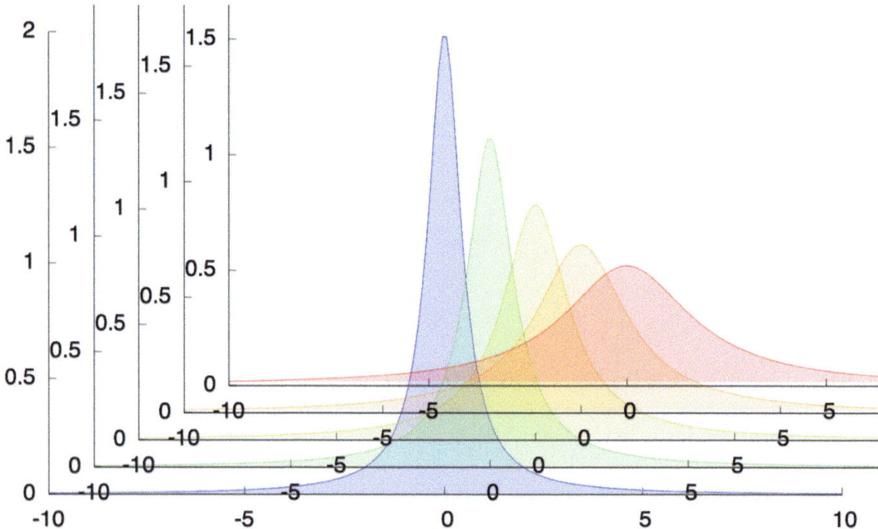

Fig. 5.46 Example of the use of 'do for' in plot command for automatic customization of plots

```
 8  set border 1+2 # plot only bottom x1 an y1 axes
 9  w= 0.5 ; dx= 0. ; dy=0.
10
11  #p lor(x,w) lc "red"
12  do for [COL in "blue green dark-yellow orange red"] {
13      p lor(x,w) w filledcurve fs transparent solid 0.2 lc rgb
        COL
14      w=w*1.4          # increase the width a bit
15      dx= dx+0.05 ; dy=dy+0.05
16      set origin dx,dy # shift the origin a bit
17  }
18  unset multi
19  reset
```

Code 5.60 Use of 'do for' iteration for cascade plot

where we have chosen the line color for each plot: `lc rgb col`. There are 111 predefined color names in gnuplot, whose list is shown through `show colornames` command, but any RGB color can be used by its hexadecimal value: `rgb '#RRGGBB'`. We have also used the `with filledcurve` style for a plot with solid transparent fill style option: `fs transparent solid 0.2`. The output of that script is shown in Fig. 5.46.

As for the `if .. then .. else` condition this is obtained by:

```
1  if (<condition>) {
2          <commands>
3          <commands>
4  } else {
5          <commands>
6  }
```

Code 5.61 Sintax of if-then-else in gnuplot

that can be used to construct also the `elseif` condition by nesting the `if` within the `else`.

```
if (<condition>) {
        <commands>
        <commands>
} else {
        if (<condition>) {
            <commands>
        }
}
```

Code 5.62 Sintax of elseif in gnuplot

The `while` condition is also supported by:

```
while (<expr>) {
     <commands>
}
```

Code 5.63 Sintax of 'while' in gnuplot

The example below shows a script to produce an animation using iteration of commands.

```
reset
set size square
unset key
set sam 50
set xrange[-25:25] ; set yrange[0:50]
a=4. ; R=1.
while (a>0) {
   p sqrt(R**2-x**2) w lp lt R ps 3
   R=R+1.
   if (R==24.) {
   R=1.
   a=a-1
   }
   pause 0.2
   }
```

Code 5.64 Example of animation using iterations

Half circles of increasing radius are plotted using all the gnuplot line styles in sequence. Here we have used the `pause <time in seconds>` command to slow down the animation, if `<time in seconds> = -1` then the execution will resume after return is pressed.

Gnuplot iteration loops may contain a `break` to exit the loop and a `continue` to skip over the remaining lines and start the next iteration, if it is still pending.

5.4.9   Solving Differential Equations

Although Gnuplot is a very versatile tool, it doesn't have built-in symbolic differential equation solvers like more specialized software. However, one can still solve differential equations numerically using it. This involves defining the differential equation and applying numerical techniques like Euler's method or Runge-Kutta, which can be implemented via GNUplot's scripts.

Considering a simple first-order differential equation, with an initial condition:

$$\frac{\mathrm{d}y}{\mathrm{d}x} = f(x, y) \qquad\qquad y(x_0) = y_0$$

We can solve this recursively, starting from the initial conditions, using Euler's Method (for simplicity):

$$y(x_0 + \Delta x) = y(x_0) + \left.\frac{\mathrm{d}y}{\mathrm{d}x}\right|_{x_0} \Delta x$$

where Δx is a small (as possible) step size.

As an example we will solve the classical projectile motion for which exists an exact solution. In the simplest case, the motion equations are:

$$\frac{\mathrm{d}v}{\mathrm{d}t} = g \qquad\qquad \frac{\mathrm{d}x}{\mathrm{d}t} = v \qquad\qquad g = -9,81$$

We can write the following GNUplot script to implement this, where we have used an implicit midpoint method for calculating the displacement.

```
 1  reset
 2  # Impostazione dei parametri
 3  dt = 0.001    # integration step
 4  kmax=10000
 5  g = -9.82
 6  # input data
 7  print 'Type inital velocity (m/s):'        # promt user for
        input
 8  V0 = system("read V0; echo $V0")           # input user choice
 9  print 'Type launch angle (deg): '          # promt user for
        input
10  theta = system("read theta; echo $theta")  # input user choice
11
12  x0 = 0.
13  y0 = 0.
14  v0x=V0*cos(theta/180.*pi)
15  v0y=V0*sin(theta/180.*pi)
16
17  !rm gravity.txt   # clean existing file output
18  set print 'gravity.txt' append\index{Append}
19  print '#  Time(sec) , X(m) , Y(m), Vx(m/s), Vy(m/s), V(m/s) '
20
```

```gnuplot
21  do for[k=1:kmax] {
22      t=k*dt
23      vy= v0y + g*dt
24      sy= (vy+v0y)/2*dt + y0
25      sx= v0x*dt + x0
26      V=sqrt(v0x**2 + vy**2)
27  # updating initial positions
28      v0y=vy
29      y0=sy
30      x0=sx
31  # checking for ground contact
32        if (y0 < 0) {
33           break    }
34  # write formatted results on file
35        print sprintf("f .10f .10f .10f .10f
        .10f", t, sx, sy, v0x, v0y, V)
36        tstop=t
37        }
38
39  set  print  # close output file and return to standard output
40  # plot session
41  set parametric  # to draw the exact solution
42  x0 = 0.
43  y0 = 0.
44  v0x=V0*cos(theta/180.*pi)
45  v0y=V0*sin(theta/180.*pi)
46  X(t)= v0x*t + x0
47  Y(t)= 1./2.*g*t**2 + v0y*t +y0
48  set trange [0:tstop]
49  set title 'Computed vs. calculated trajectory   V0= '.V0.'m/s
         --   Theta= '.theta.'deg' # title will report used input
         value
50  p 'gravity.txt' us 2:3 every 20 w p pt 4 t 'Computed', X(t),Y(
         t) w l t 'Calculated'  # every: skips points
51  unset title
52  set term x11 2          # open another plot window
53  set size ratio 0.2      # set plot height/width ratio
54  set ytics  auto 2e-11   # fix tics increment
55  # draw the difference between computed and exact solution
56  p 'gravity.txt' us 2:($3-Y($1)) ev 10 w i # mixed columns
         operations
57  unset parametric
58  set term x11 1  # restore first plot window
```

Code 5.65 Example of the use of Euler's Method in gnuplot

The graphics produced by that script are shown in Fig. 5.47. The top plot compares the computed and exact solutions, while the bottom plot shows their differences with impulses. We have also used the `every` qualifier to reduce the number of plotted points that otherwise would have produced a clumsy figure.

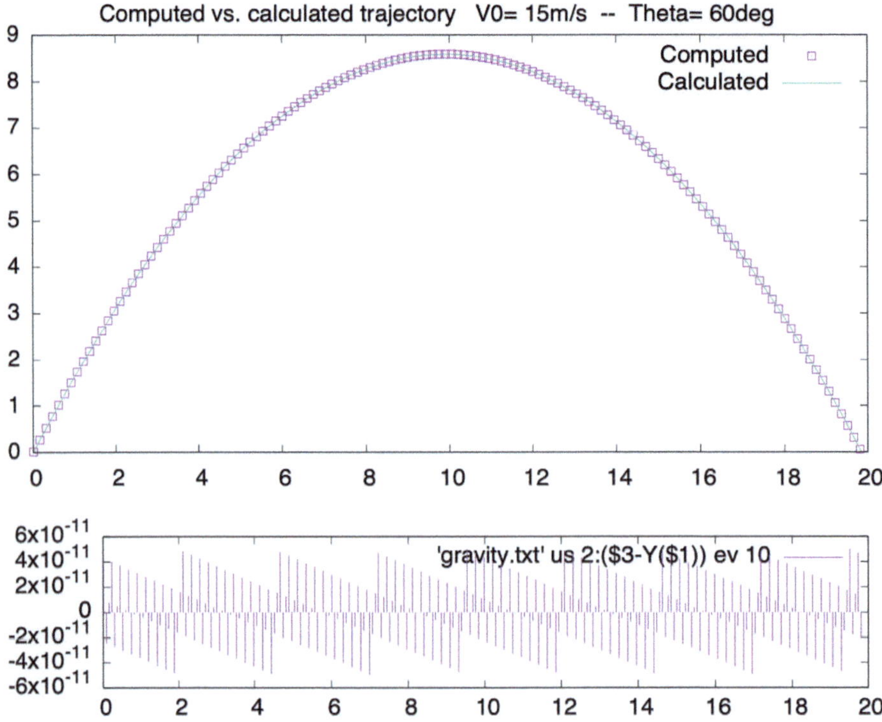

Fig. 5.47 Graphical outputs produced by the previous script. The top graph shows the computed (points) and exact (line) trajectories, while the bottom graph shows their differences

5.4.10 Statistics

As for the statistical analysis of distributions of data (i.e. single column data) gnuplot offers a simple solution, by means of the built-in routine `stats`

```
stats {<ranges>} 'filename'  using N{:M} {name 'prefix'}
```

Code 5.66 Sintax of 'stat' command

which provides a printed output of a statistical summary of the data found in the chosen columns of 'filename'. Data points are filtered against both xrange and yrange. The output can be redirected to a file by the `set print 'outfile'` command or suppressed by adding the `nooutput` option. In any case, the command stores the individual statistics into three variables.

The first set of variables concerns the set of analysed data:

```
STATS_records      # total num. of in-range data records (N)
STATS_outofrange   # num. of records filtered out by range
       limits
STATS_invalid      # num. of invalid/incomplete/missing records
STATS_blank        # num. of blank lines in the file
```

```
 5  STATS_blocks        # number of indexable blocks of data in the
        file
 6  STATS_columns       # num. of data columns in the first row of
        data
```

Code 5.67 Global variables evaluated by the 'stat' command

The properties of the in-range data from a single column are reported in the second set of variables. In case of two columns analysed together in a single command, then suffix "_x" or "_y" is appended to variable names (i.e. STATS_min_x and STATS_min_y). In this case, points are filtered against both xrange and yrange.

```
 1  STATS_min           # minimum value of in-range data points
 2  STATS_max           # maximum value of in-range data points
 3  STATS_index_min     # index i for which data[i] == STATS_min
 4  STATS_index_max     # index i for which data[i] == STATS_max
 5  STATS_lo_quartile   # value of the lower (1st) quartile
        boundary
 6  STATS_median        # median value
 7  STATS_up_quartile   # value of the upper (3rd) quartile
        boundary
 8  STATS_mean          # mean value of the in-range data points
 9  STATS_ssd           # sample standard deviation (in-range data)
10  STATS_stddev        # population standard dev. (in-range data)
11  STATS_sum           # sum
12  STATS_sumsq         # sum of squares
13  STATS_skewness      # skewness of the in-range data points
14  STATS_kurtosis      # kurtosis of the in-range data points
15  STATS_adev          # mean absolute deviation (in-range data)
16  STATS_mean_err      # standard error of the mean value
17  STATS_stddev_err    # standard error of the standard deviation
18  STATS_skewness_err  # standard error of the skewness
19  STATS_kurtosis_err  # standard error of the kurtosis
```

Code 5.68 Variables evaluated by the 'stat' command, for a chosen coulumn

All this information is also helpful in a script that automates the analysis and plotting of several data files.

The third set of variables is only relevant to cross-analyzing two data columns.

```
 1  STATS_correlation   # correlation coefficient between x and y
 2  STATS_slope         # A corresponding to a linear fit y = Ax +
        B
 3  STATS_slope_err     # uncertainty of A
 4  STATS_intercept     # B corresponding to a linear fit y = Ax +
        B
 5  STATS_intercept_err # uncertainty of B
 6  STATS_sumxy         # sum of x*y
 7  STATS_pos_min_y     # x coordinate of a point with minimum y
 8  STATS_pos_max_y     # x coordinate of a point with maximum y
```

Code 5.69 Variables evaluated by the 'stat' command, for cross correlation

The name option in the stats causes the default prefix "STATS" to be replaced by the specified string. It is also possible to use the analyzed column's header (first line) os a name.

```
1 stats 'datafile' using N name 'whatyoulike'  # N = column
    number
2 stats 'datafile' using N name columnheader
3 do for [COL=5:8] {stats 'datafile' using COL name columnheader
    }
```

Code 5.70 Example for assigning custom name for the output of 'stat' command

5.4.11 Histograms

Gnuplot allows a specific style to plot a histogram: `set style data histogram`. This produces a bar chart from a sequence of parallel data columns. In the plot command must be specified a single column of the input file, possibly with related axis tic values or key (legend) titles.

For example, we can consider the following file containing the lifetime distribution for a set of incandescent bulb lamps and neon tubes.

```
1  #lifetime(hours) neon bulb
2  500    ,    , 35
3  1000   ,    , 80
4  1500   ,    , 70
5  2000   ,    , 48
6  3000   , 14
7  4000   , 56
8  5000   , 60
9  6000   , 86
10 7000   , 74
11 8000   , 62
12 9000   , 48
```

Code 5.71 Example of file containing statistical data on lamp liftime

This data can be easily plotted as a histogram by the following lines:

```
1 reset
2 set datafile separator ','  # since we have comma separated
    values
3 set style data histogram
4 set boxwidth 3      # default is 2
5 set style fill solid border -1  # -1 is a solid line on most
    terminal
6 set xlabel 'Lifetime (hours)'  font 'arial,20'
7 set ylabel 'Tested lamps'  font 'arial,20'
8 plot 'Histo-Lamp.txt' u 2:xtic(1) t 'NEON' lc 'blue','' u 3 t
    'BULB' lc 'red'
```

Code 5.72 List of command to produce an histogram

obtaining the graph in Fig. 5.48 where we have used column 1 for tics on x axes by the specification: `using 2:xtic(1)`.

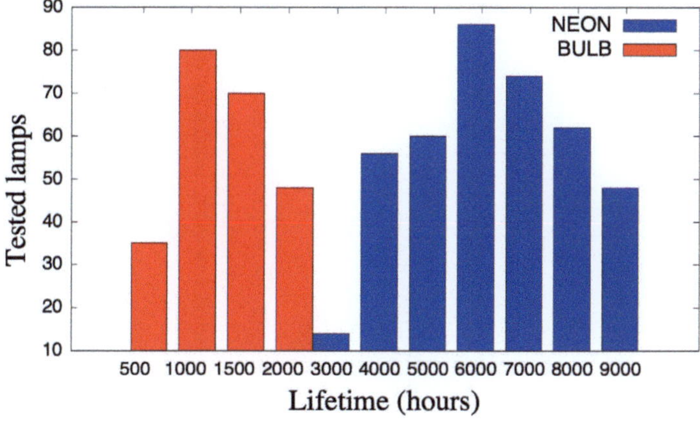

Fig. 5.48 Histogram of lifetime distribution for incandescent bulb lamps and for neon tubes

As another example where the class of histogram are text strings rather than numbers, as it is usual in histograms, is given in the following file reporting the average height per gender in European countries:

```
#Country Male    Female  gender
Holland 183     170     male
Sweden  181     167     female
Norway  181     167
Denmark 181     169
Spain   175     162
France  176     162
Germany 180     166
Italy   175     161
UK      177     163
Greece  178     165
```

Code 5.73 Example of file containing statistical data on average height per gender in European countries

In this case, the following set of commands will plot the histogram

```
reset
set style data histogram
set style fill solid border -1
set style histogram cluster gap 2 # gap is optional
set yr [0:]
set xlabel 'Country'  font 'arial,20'
set ylabel 'Average Height (cm)'  font 'arial,20'
plot 'Histo-Height.txt' u 2:xtic(1) t 'Male' lc 'skyblue','' u
     3 t 'Female' lc 'pink'
```

Code 5.74 List of command to produce another histogram

that it is shown in Fig. 5.49.

Furthermore, it is also possible to stack the histogram boxes by row or by column, as in the following example

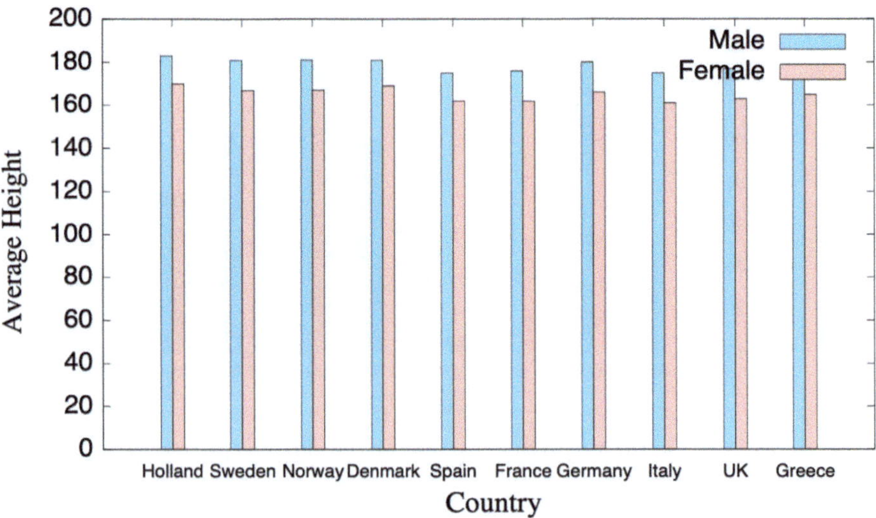

Fig. 5.49 Histograms of the average height per gender in European countries

```
1  reset
2  set style data histogram
3  set boxwidth 0.9 relative # to leave a small gap among boxes
4  set key out
5  set style fill solid border -1
6  set style histogram cluster
7  set style histogram rowstacked # ROWS AWARE
8  unset ytic
9  plot 'Histo-Height.txt' u 2:xtic(1) t 'Male', '' u 3 t 'Female
      '
```

Code 5.75 List of command to produce stacked histograms

and

```
1  reset
2  set style data histogram
3  set boxwidth 0.9 relative # to leave a small gap among boxes
4  set style fill solid border -1
5  set style histogram cluster
6  set style histogram columnstacked # COLUMNS AWARE
7  unset ytic
8  plot 'Histo-Height.txt' using 2:key(1), '' using 3:xtic(4) #
      note the use of key and xtic
```

Code 5.76 List of command to produce piled histograms

In this case, the outputs are shown in Fig. 5.50 where ytics has been removed because it is no longer informative.

We can now consider the use of `stats` command to perform some operation on the data in order to produce a graph like that in Fig. 5.49, but more informative.

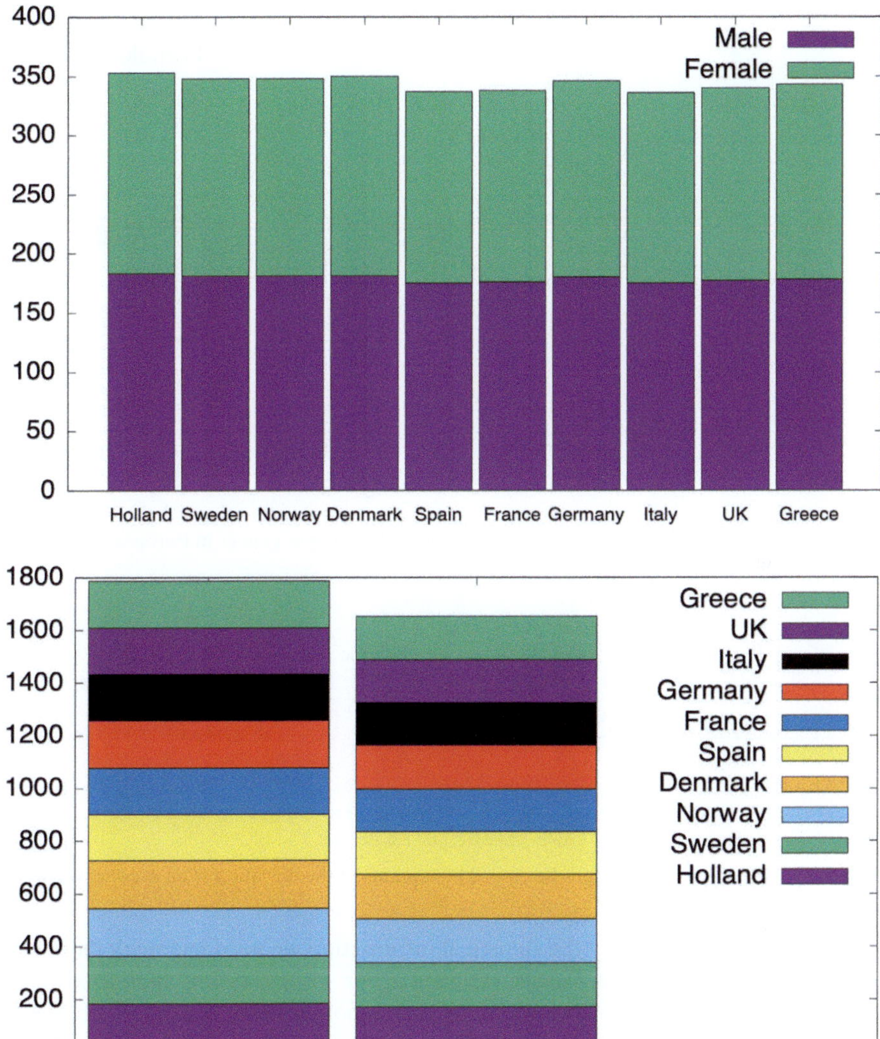

Fig. 5.50 Example of piled and stacked histogram styles

In particular, we can plot the deviation from average values for the two genders by extracting the average values from `stats` and plotting the difference on the fly.

```
reset
stats 'Histo-Height.txt' u 2 name 'M' noout # nooutput quiet
    mode
stats 'Histo-Height.txt' u 3 name 'F' noout
set style data histogram
```

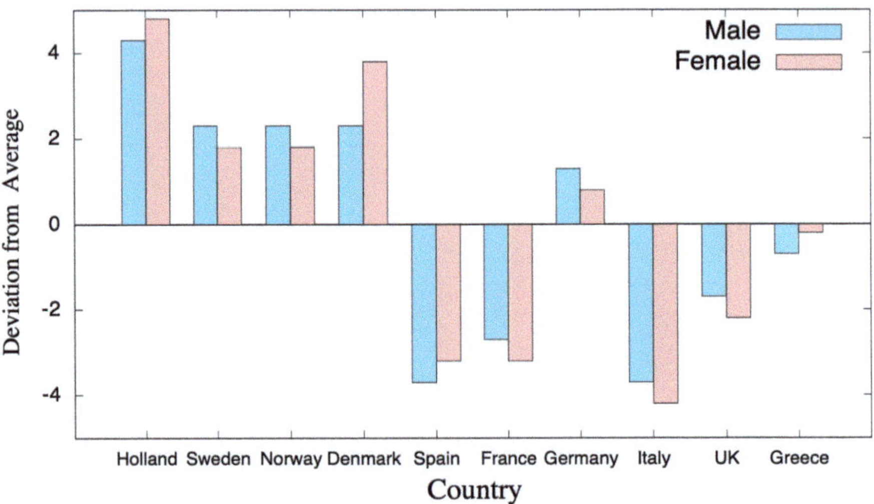

Fig. 5.51 Histogram of the deviance from the average height, per gender in European countries

```
5  set style fill solid border -1
6  set style histogram cluster gap 1
7  set yr [-5:5] ; set xzeroaxis lt -1 lw 2
8  set xlabel 'Country'  font 'arial,20'
9  set ylabel 'Deviation from Average Height'  font 'arial,20'
10 Mtit=sprintf('Male Average height= %3.1f ',M\_mean) #title
       string
11 Ftit=sprintf('Female Average height= %3.1f ',F\_mean)
12 plot 'Histo-Height.txt' u ($2-M_mean):xtic(1) t Mtit lc '
       skyblue', '' u ($3-F_mean) t Ftit lc 'pink'
```

Code 5.77 List of command to produce piled histograms of the deviation from average values for both gender

The first two lines extract the statistical information we need and mark them with a different prefix, the mean values (M_mean) or (F_mean) is then used in the plot command ($2-M_mean) and in the key reported in the legend, to obtain the plot in Fig. 5.51.

Until now, we have dealt with data already aggregated in histogram form. We will now see how to create a histogram from an un-aggregated distribution of data, as shown in Fig. 5.52. In order to make a histogram out of raw data, we have to write a function that clusters data points in a chosen set of bins (intervals on the x-axis), and then we have to count how many points there are in each interval. The first part of the job is done by an user-defined function, while the second part is implemented by the built-in smooth frequency in gnuplot, as illustrated in the script below:

```
1 reset
2 binwidth = 5
3 bin(x,width) = width * floor(x / width)
4 set boxwidth binwidth      # Match boxwidth to binwidth
```

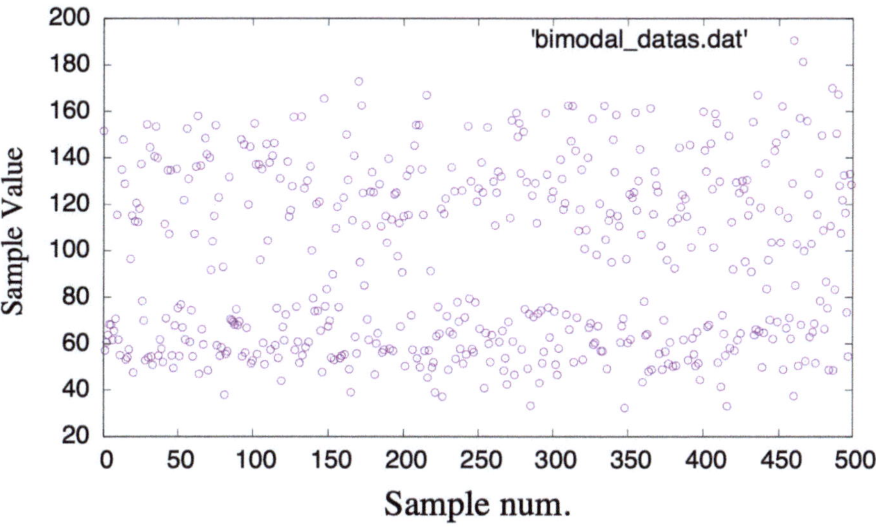

Fig. 5.52 Example of an un-aggregated distribution of data

```
5  set style fill solid 0.5   # Half transparent filling
6  set xlabel "Data Value" font 'arial,20'
7  set ylabel "Frequency"  font 'arial,20'
8  plot 'bimodal_datas.dat' using (bin($1,binwidth)) smooth freq
      with boxes
```

Code 5.78 Script to create and plot histogram from unaggregated data

The `bin(x,width)` function round the data in bins of chosen width, then the plot option `smooth frequency` makes the data monotonic in x and points with the same x-value are replaced by a single point having the summed y-values, thus building up the histogram. The plot style `with boxes` just makes up the vertical bars, but also points or lines can be used. The resulting plot is shown in Fig. 5.53.

The obtained data can also be saved to a file and used for further analysis. As told, this is achieved by `set table 'filename'`, repeating the last plot command, and `unset table`. In our case, the obtained file is:

```
1  # Curve 0 of 1, 31 points
2  # Curve title: "'bimodal_datas.dat' using (bin($1,binwidth))"
3  # x y xlow xhigh type
4  30   3   30   30   i
5  35   5   35   35   i
6  40   7   40   40   i
7  45   23  45   45   i
8  50   44  50   50   i
9  55   42  55   55   i
10 60   41  60   60   i
11 65   38  65   65   i
12 70   26  70   70   i
13 75   16  75   75   i
```

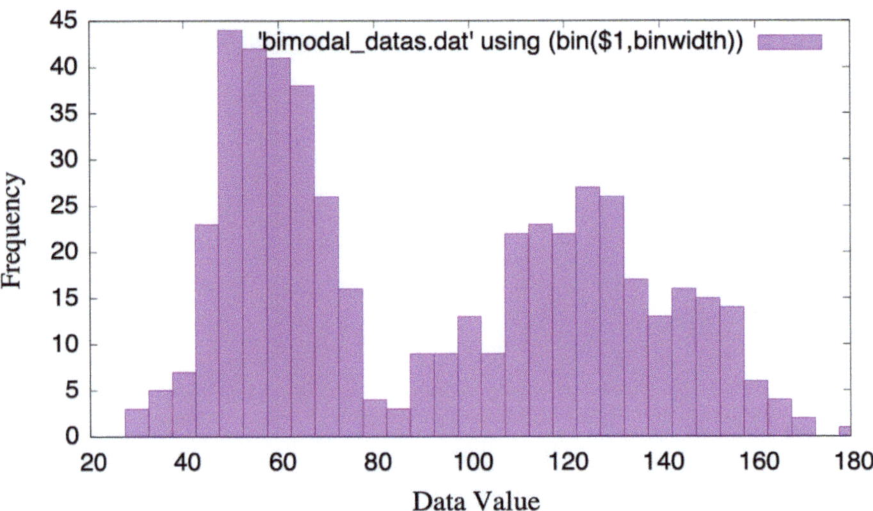

Fig. 5.53 Histogram obtained by the un-aggregated distribution of data

```
14   80   4   80   80   i
15   85   3   85   85   i
16   90   9   . . . . . . . . .
17   . . . . . . . . . . . . . . . .
```

Code 5.79 Example of printed output of the binned data

5.4.12   3D Plot

Gnuplot can also be used to display surfaces (2D arrays) using the `splot` command instead of `plot`. As a simple example, we can draw a bidimensional sinc function $(\sin(R)/R)$:

```
1   reset
2   unset key
3   set surface  # default for splot
4   set isosam 50,50       # equivalent of 'set sample' in 1D
5   R(x,y) = sqrt(x**2 + y**2)
6   splot sin(R(x,y))/R(x,y)
```

Code 5.80 Simple script for a 3D plot of sin(r)/r function

obtaining the plot in Fig. 5.54 left pane. The plot can also be colored with a palette (variable colors with z values) and rendered as non-transparent, moreover, we can add a contour plot as shown in the right pane, by using:

```
1   set contour    # add contour plot at bottom
2   set hidden3d  #   set hid for short, to make solid figure
```

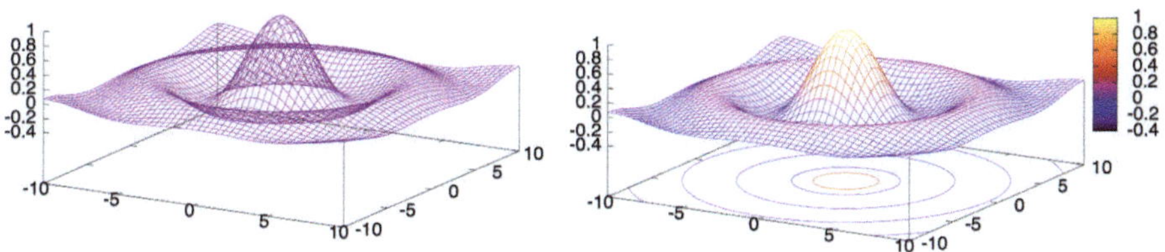

Fig. 5.54 Surface plots of sinc(x) function in monochrome and color gradient style

```
3  R(x,y) = sqrt(x**2 + y**2)
4  splot sin(R(x,y))/R(x,y) lc pal # line color palette
```

Code 5.81 Script for add a contour plot and a colour palette to a 3D plot

The point of view from which the 3D object is observed is controlled by the `set view rotation, elevation` command; the default values are rotation = 60°, elevation = 30° (the x-rotation is performed first).

When an interactive terminal, like X11 or wxt, is used it is also possible to pan and zoom the graph by mouse gesture (very satisfying on fast computers) and then visually find the optimal point of view to be used for printout using pdf/png/jpg terminals, by setting the values grabbed with `show view` command in the terminal.

Of course, line color, thickness, and palette gradient can be highly customized, use the in-line `help palette` utility for details.

It is also possible to plot, for smoother shading, an interpolated continuous surface by specifying `set pm3d interpolate N,N` (where `N` is an integer to increase the interpolation). It uses an algorithm that allows plotting gridded as well as non-gridded data without preprocessing, even when the data scans do not have the same number of points.

This setting can be used globally by `set pm3d` or just for a single graph: `splot ...... with pm3d`. In the former case, the pm3d surface is drawn along with the mesh produced by the style specified in the splot command, as shown in Fig. 5.55 left and right panes.

```
1  reset
2  set pm3d
3  set iso 20,20
4  R(x,y) = sqrt(x**2 + y**2)
5  splot sin(R(x,y))/R(x,y)   w l lw 2 lc 'black'
6  pause -1 'Click to continue'
7  unset pm3d
8  set iso 100,100
9  splot sin(R(x,y))/R(x,y)   w pm3d
```

Code 5.82 Script showing the use of pm3d to plot an interpolated surface

The use of `pm3d` allows for producing maps, that are actually the projection of a 3D object on the xy plane. This can be attained by `set view 0,90`, i.e., looking at the object from the top.

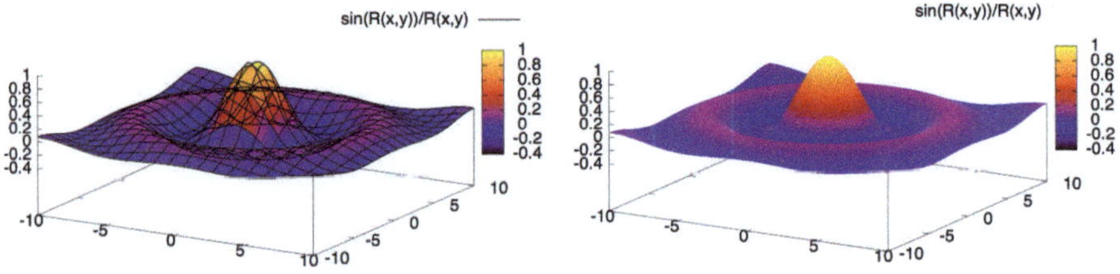

Fig. 5.55 Example of interpolated surfaces with the use of 'pm3d' options

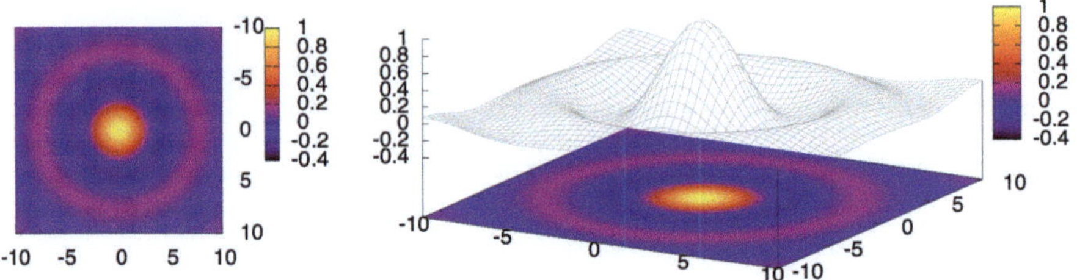

Fig. 5.56 Examples of the use of 'pm3d' to plot bidimensional maps from surfaces

```
reset
set view equal #   make a square graph window
set view 0,90  #   for a top down projection
unset key
set iso 100,100
R(x,y) = sqrt(x**2 + y**2)
splot sin(R(x,y))/R(x,y)  w pm3d
```

Code 5.83 Script showing the use of pm3d to plot a top down projection of a surface

This will produce the left graph in Fig. 5.56.

The same results can be obtained by using `with pm3d at bottom` that is not dependent on the view setting.

This behavior can be made permanent by `set pm3d map` that allows the use of lines (or any other style) to `splot` the surface, as shown on the right of the same figure.

```
reset
set pm3d at b
set iso 50,50
set hidden
R(x,y) = sqrt(x**2 + y**2)
splot sin(R(x,y))/R(x,y)  lc 'dark-gray' not
```

Code 5.84 Script showing the combined use of pm3d and line/point style

Of course, what has been said about functions also applies to a 3D plot of user-tabulated data. Gnuplot builds up surfaces from data points that are organized in set of three columns as x, y, z, arranged into "blocks" separated by a blank line (similar

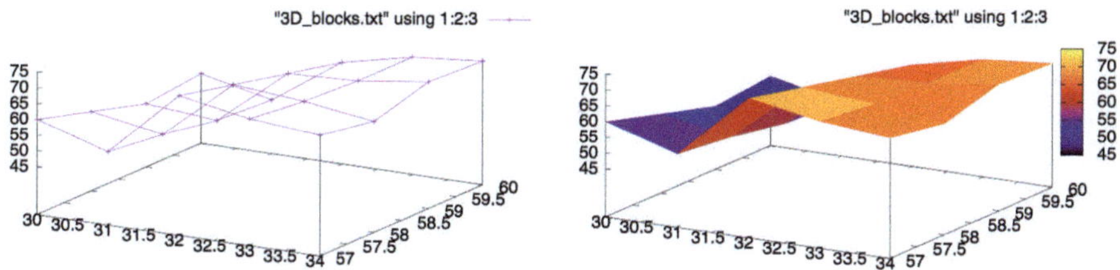

Fig. 5.57 Example of surface plot from a 3D user blocks data file

to 2D dataset). In each block the x column is repeated while the y and z are vary from block to block, like in the example below (3D_blocks.txt):

```
30   60   52
31   60   47
32   60   62
33   60   67
34   60   69

30   59   50
31   59   48
32   59   66
33   59   67
34   59   70

30   58   55
31   58   51
32   58   70
33   58   68
34   58   65

30   57   60
31   57   53
32   57   74
33   57   70
34   57   68
```

Code 5.85 Example of a 3D user data file in block format

That actually consists in a set of slices of the surfaces with planes normal to the x axis. This block file can be plotted by:

```
reset
set pm3d
splot "3D_blocks.txt" using 1:2:3 with lp   # lines and points
# Or for palette colored surface with:
splot "3D_blocks.txt" using 1:2:3 with pm3d   # filled surface
```

Code 5.86 Example of scripts to plot from a 3D user data file

obtaining the graphs in Fig. 5.57.

The option `pm3d` has to be set before to have a filled surface and lines grid together. For smoother shading, use `set pm3d interpolate N, N` where `N` is an integer that controls the interpolation grid; the default value is 1, 1.

Alternatively, it is possible to use a regular `matrix` data format where the rows and columns indices represent the grid of `x` and `y` values, and each entry is a `z` value.

```
z[1,1]  z[1,2]  z[1,3]
z[2,1]  z[2,2]  z[2,3]
z[3,1]  z[3,2]  z[3,3]
```

Code 5.87 'Regular matrix' example of formatted 3D user data file

Otherwise, one can deserve the first column and the first row for `x` and `y` coordinates by specifying `matrix non uniform` as in the example below:

```
0       20      60      100     140     180     220     260
. . . . .
0.20    1.000   0.944   0.865   0.847   0.760   0.701   0.686
. . . . .
0.52    0.998   0.933   0.866   0.811   0.780   0.685   0.609
. . . . .
0.78    0.996   0.930   0.864   0.803   0.732   0.673   0.548
. . . . .
1.27    0.990   0.902   0.875   0.794   0.698   0.592   0.383
. . . . .
1.47    0.987   0.921   0.867   0.766   0.665   0.549   0.316
. . . . .
1.67    0.983   0.907   0.828   0.773   0.613   0.488   0.242
. . . . .
1.87    0.979   0.880   0.833   0.742   0.608   0.469   0.198
. . . . .
2.06    0.975   0.900   0.823   0.758   0.560   0.414   0.165
. . . . .
2.27    0.970   0.857   0.846   0.730   0.512   0.376   0.129
. . . . .
2.46    0.964   0.863   0.810   0.712   0.500   0.359   0.116
. . . . .
2.65    0.959   0.850   0.834   0.721   0.513   0.371   0.151
. . . . .
2.83    0.953   0.829   0.787   0.648   0.412   0.267   0.067
. . . . .
. . . .  . . . . .  . . . . .  . . . . .  . . . . .  . . . . .  . . . . .  . . . . .
. . . . .
```

Code 5.88 'Non uniform matrix' example of formatted 3D user data file

The output from `matrix` type data and from `matrix nonuniform` type it is shown in Fig. 5.58 and can be obtained respectively by:

```
splot '3D_Matrix_1.txt' matrix w pm3d
splot '3D_Matrix_2.txt' matrix nonuniform w pm3d
```

Code 5.89 Simple scrit to plot 'regular' and 'non uniform' matrix from user data

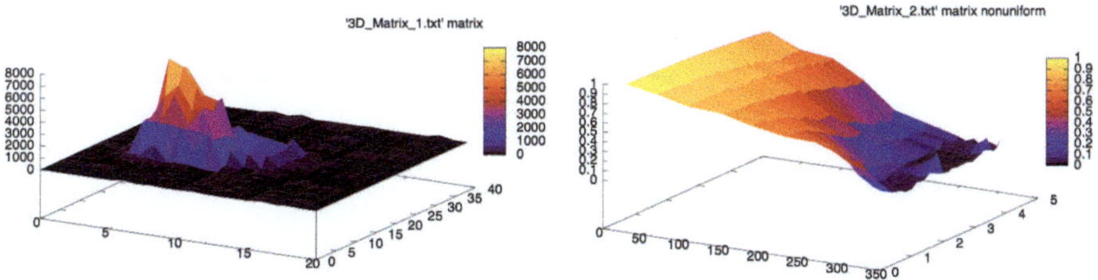

Fig. 5.58 Example of surface plots for 'matrix' and for 'matrix nonuniform' data type

5.4.13 Miscellanea

Axis
The four available axes are:
'x1y1' axes on the bottom and left;
'x2y2' axes on the top and right;
'x1y2' axes on the bottom and right;
'x2y1' axes on the top and left.
They are selected by `axes ##` in the plot command to associate a curve to the selected axes. Ranges specified on the 'plot'command apply only to x1y1.

In the example below, we will have a graph with double y axes with different colors:

```
reset
set xrange [-10:10]
set ytics 10 nomirror tc lt 1 # tic color is that of line type
set ylabel 'Quadratic'
set y2tics 200 nomirror  tc 'blue' # tic color user defined
set y2label 'Cubic'
plot x**2 linetype 1, x**3  lc 'blue' axes x1y2
```

Code 5.90 Simple scrit illustrating the use of 'axes' specification

where for the left y axes the color is associated with the line type, while for the right one it is chosen by the user, in both cases, the same association is to be repeated in the plot command. The output is shown in Fig. 5.59.

In the case a logarithmic scale is needed, it can be activated by `set log 'axis' 'base'` where 'axis' is any combination of `x, y, x2, y2`, while default is base 10.

It is also possible to use polar coordinates by `set polar`. In this case the dummy variable (t) represents an angle theta with default range of $[0:2*\pi]$.

Datablock

Datablocks in gnuplot are a kind of virtual files that aren't saved on mass storage and are used as online data or embedded in scripts. A $ prefix identifies Datablocks names and can be deleted using `undefine` command, while `undefine $*` frees all data blocks.

Fig. 5.59 Example of plots with different scales on the axis

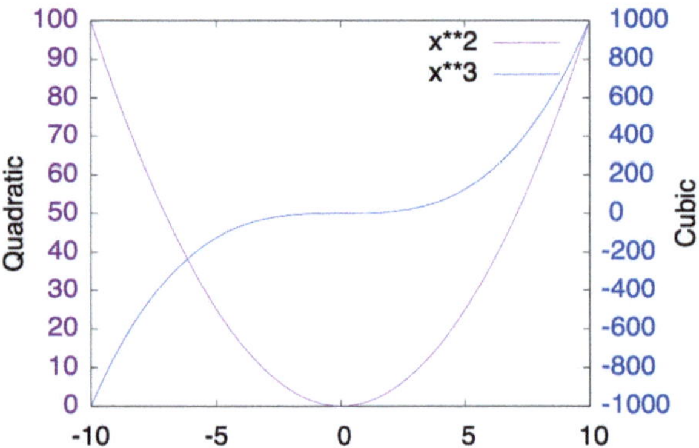

A datablock can be defined as in this example:

```
# Define a data block
gnuplot> $Mydata << EOD
11 22 33 #first line of data
44 55 66 #second line of data
# comments work just as in a data file
77 88 99
EOD
gnuplot>print $Mydata
11 22 33
44 55 66
77 88 99
gnuplot> p $MyData us 1:2 w lp
```

Code 5.91 Example to illustrate the creation of a 'data block' within the gnuplot terminal

The end-of-data delimiter (EOD in the example) may be any sequence of alphanumeric characters. Once a data block has been defined, it can be printed, plotted, manipulated, or fitted as any ordinary data file.

The Datablocks function allows for encapsulating multiple data blocks inside a script without needing to use external files, making scripts more portable.

Datablocks are also useful for temporary or intermediate datasets, such as intermediate steps in data processing. The user can define and reuse these datasets without cluttering the filesystem with temporary files.

The special filename '-' is another mechanism for embedding data into a stream of gnuplot commands. When this appears in a plot command, the lines immediately following the plot command are interpreted as inline data, but they can only be used once by the following plot command.

```
gnuplot> plot '-'
input data ('e' ends) > 1 2
input data ('e' ends) > 3 4
input data ('e' ends) > 5 6
input data ('e' ends) > 7 8
input data ('e' ends) > 9 10
input data ('e' ends) > e
gnuplot>
```

Code 5.92 Anoher example to illustrate the creation of a data file within the gnuplot terminal, using plot '–'

In this example, the user only needs to input the data at the prompt `input data ('e' ends) >` while the plot will be produced just after entering the `'e'` character.

Coordinates

To draw something like an arrow, key, label, etc., at an arbitrary position in a plot window, it is necessary to specify a reference system and the position in that system. This is achieved by the following syntax:

`{system} x, {system} y, {system} z}`

wher `system` can either be first, second, polar, graph, screen or character, according to the following rules.

first   x, y, or z coordinate is defined by the values of the left and bottom axes
second   the same as first but referred to the x2, y2 axes (top and right)
graph   specifies the area within the axes: 0,0 is bottom left and 1,1 is top right (for splot, 0, 0, 0 is bottom left of the plotting area
screen   specifies the whole screen area, with 0, 0 at bottom left and 1, 1 at top right.
character   used primarily for offsets, not absolute positions. The character size depends on the used font
polar   causes the first two values to be interpreted as angle theta and radius r

The default coordinate system for x is `first`, if the system for y is not specified, then the specification for x is assumed.

Smoothing

Gnuplot includes a general-purpose data interpolation routine under the 'smooth' qualifier in plot command: `plot 'filename' using n:m smooth 'option'` The most commonly used options that construct a continuous curve between a set of data are:

csplines: connects consecutive points by natural cubic splines fitted to data points, after rendering the data monotonic in x.
acsplines: same as cspline but with the coefficients of cubic polynomials that are weighted by the error on the points, indicated by the third column in `plot 'filename' us X:Y:Err`
bezier: approximates the data with a Bezier curve of degree equal to the number of data points.

Other available options are useful for statistical analysis. In fact, after sorting the data on x, this functions plot some aspect of the distribution of x values.

unique: points with the same x-value are replaced by a single point having the average y-value.
frequency: points with the same x-value are replaced by a single point having the summed y-values.
cumulative: points with the same x-value are replaced by a single point containing the cumulative sum of y-values of all data points with lower x-values.

Color palette

Palette or color gradient, used in surface plot, can be customized to better represent the data. A *palette* in Gnuplot is a set of colors used to represent data in color plots, such as heat maps or 3D surface plots. By defining a palette, the user can control how numerical values are translated into colors, making it easier to visualize data distributions and z intensity variations. The palette legend can be removed by the `unset colorbox` command if necessary.

The simple command `set palette` will activate the default blue-to-red gradient palette. Otherwise, it is possible to define a custom palette by coupling intensity values to specific colors

```
1 reset
2 set palette defined (0 "blue", 0.5 "white", 1 "red")
```

Code 5.93 Example of used defined palette

In this case, the numerical values (0, 0.5, 1) indicate the minimum, halfway, and maximum intensities in the data file or function values in the plotted range. In this example, values close to 0 will be blue, around 0.5 will be white, and near 1 will be red. In the palette definition, the'color switch' points are not limited to three but can be increased at will. `set palette defined (value1 "color1", value2 "color2", ..., valueN "colorN")`, providing fine tuning of color gradients.

As told, gnuplot will generate a gradient using the defined color and switch points, but it is also possible to have discrete colors by using `set palette maxcolors 'num_of_colors'` Use the command `show palette colornames` for a list of available color names. The palette definition can also be saved and loaded from a text file that has the following format:

```
0.0 0.0 0.0 1.0
0.5 1.0 1.0 1.0
1.0 1.0 0.0 0.0
```

each line contains the 'switch point' and the assigned color as a combination of Red, Green, and Blue. Those files are loaded via `set palette file 'palette.txt'`.

Gnuplot also offers a set of predefined palettes that are accessible via `set palette rgbformulae 33, 13, 10`, where the used triple is just gnuplot indexes that refer to formulae for the red, green, and blue components. As an example of a commonly used palette, we have:

7, 5, 15 for Classic rainbow

33, 13, 10 for Viridis-style gradient

34, 35, 36 for Plasma-style gradient

3, 3, 3 for Grayscale

For a complete list of all possible preset combination type `show palette rgbformulae` in gnuplot terminal.

Users can also define a palette using mathematical functions using the `set palette functions <R>, <G>, <B>` command. Where `<R>`, `<G>`, and `<B>` are expressions that determine the red, green, and blue components, as for example `set palette functions sin(0.3*pi*gray), gray**2, 1-gray`

It is even possible to use a palette to plot a 2d file using a third column that determines the color of each point: `plot "datafile.dat" using 1:2:3 with points palette`

5.5 Data Manipulation and Visualization with Python

Let's now dive into the data manipulation and visualization using Python. As already stated, Python offers a rich ecosystem of libraries that enhance its capabilities across various domains. For data manipulation, Python libraries like "NumPy"—numerical computing, array manipulation, and linear algebra –, "SciPy"—for scientific and technical computing (optimization, integration, etc.)—and "Pandas"—database manipulation, analysis, and cleaning—are commonly used. Concerning data visualization, Python offers the use of several libraries such as "Matplotlib"—for creating static, interactive, and animated plots.

In the following sections, we will see specifically what these Python libraries consist of and why they are so powerful.

5.5.1 *NumPy*

NumPy (short for Numerical Python) is a fundamental package for scientific computing in Python. It offers an extensive set of routines/functions to perform fast operations and complex calculations without manual iteration:

- Mathematical functions (e.g., element-wise addition, multiplication);
- Logical operations (e.g., masking, boolean indexing);
- Shape manipulation (e.g., reshaping, transposing);
- Sorting, selecting, and input/output (I/O);
- Basic linear algebra (matrix operations);
- Basic statistical operations;
- Discrete Fourier transforms;
- Random simulation (random number generation).

NumPy is efficient and performs very well. In fact, NumPy operations are often performed in "compiled code," which significantly improves execution speed. NumPy outperforms Python's built-in sequences (like lists) for various scientific computing tasks. For these reasons, many scientific and mathematical Python-based packages rely on NumPy arrays. These packages often convert Python sequences to NumPy arrays for efficient processing. It is well evident that NumPy is being used extensively and knowing how to use NumPy is essential for working with scientific software.

NumPy allows you to create and manipulate arrays of homogeneous data types (e.g., integers, floats) efficiently by means of the "array" and "ndarray" (n-dimensional array) objects. Unlike Python lists, NumPy arrays have a fixed size at creation, which leads to better performance and allows you to handle large amounts of data quickly and flexibly. Arrays can be initialized with NumPy using a Python sequence, such as a list. After installing NumPy (e.g., via "pip"), NumPy may be imported into Python code as reported in Line 1 of Code 5.94.[2]

```python
import numpy as np

a = np.array([2, 4, 6, 8, 10])
a
# Output: array([2, 4, 6, 8, 10])
print(a)
# Output: [2 4 6 8 10]
```

Code 5.94 Simple array with data in NumPy

As with built-in Python sequences, NumPy arrays are "0-indexed": the first element of the array is accessed using index 0, not 1. Like the original list, array elements can be accessed using the integer index of the element within square brackets, and Python slice notation can be used for indexing. Furthermore, arrays are mutable.

```python
import numpy as np
a = np.array([2, 4, 6, 8, 10])
#Indexing
a[0]
# Output: 2
a[:3]
# Output: array([2, 4, 6])
#Variability
a[0] = 11
a
# Output: array([11, 4, 6, 8, 10])
```

Code 5.95 Indexing and variability of arrays in NumPy

Slicing an array returns an object that refers to the data in the original array—a so-called *view*.[3] So, it is possible to mutate the original array using the view, as shown in the following example:

```python
import numpy as np

a = np.array([2, 4, 6, 8, 10])
```

---

[2] Note that (import numpy as np) is only a common convention that allows the access to NumPy features with a short and recognizable prefix (np).

[3] Refer to https://numpy.org/doc/stable/user/basics.copies.html for more info about views and the their differences with *copies* for a more comprehensive explanation of when array operations return views rather than copies.

```
4
5  #Slicing
6  v = a[2:]
7  v
8  # Output: array([6, 8, 10])
9  v[0] = 1
10 a
11 # Output: array([ 2,  4,  1, 8, 10])
```

Code 5.96 Slicing and views in NumPy

So far, we have dealt only with one-dimensional (1D) arrays. By using nested Python sequences, we can obtain two- and higher-dimensional arrays, getting the number of dimensions, shape—length of each dimension—, and size—total number of elements—of a NumPy array using the ndim , shape , and size attributes, respectively.

```
1  import numpy as np
2
3  a = np.array([[2, 4, 6, 8], [10, 12, 14, 16], [18, 20, 22,
       24]])
4  a
5  # Output: array([[ 2,  4,  6,  8],
6  #                 [ 10, 12, 14, 16],
7  #                 [ 18, 20, 22, 24]])
8  print(a)
9  # Output: [[ 2  4  6  8]
10 #          [10 12 14 16]
11 #          [18 20 22 24]]
12
13 a.ndim
14 # Output: 2
15
16 print(a.shape)
17 # Output: (3, 4)
18 print(type(a.shape))
19 # Output: <class 'tuple'>
20 print(a.shape[0])
21 # Output: 3
22 print(a.shape[1])
23 # Output: 4
24 row, col = a.shape
25 print(row)
26 # Output: 3
27 print(col)
28 # Output: 4
29
30 print(a.size)
31 # Output: 12
32 print(type(a.size))
33 # Output: <class 'int'>
34 print(a[0].size)
35 # Output: 4
```

Code 5.97 Higher-dimensional arrays in NumPy and their characteristics

In Example 5.97, the array a is a 2D-array (a matrix) with 12 elements grouped into a matrix of 3 rows and 4 columns. Note that it has a dimension equal to 2 because it consists of two indices, one for rows and the other for columns. Graphically, the array can be represented in two-dimensional space, that is, on the (x, y) plane. Generally speaking, in NumPy, a dimension of an array is linked to an "axis". This expression makes it possible to distinguish appropriately between the dimensionality of an array and the dimensionality of the data represented by the array. Furthermore, the size should not be confused with the number of elements in the array or the number of rows and columns. In fact, the number of elements is defined by using the `size` method; instead, the number of rows and columns can be found by means of the `shape` method.

NumPy is really useful for doing data analysis in physics.[13] Below is reported an example of measuring the falling times of a body and the subsequent calculation of the "mean" of the linked dataset:

```python
import numpy as np

# Example data (fall times of an object)
fall_times = np.array([1.2, 1.5, 1.8, 1.4, 1.6])

# Calculate the mean
mean_fall_time = np.mean(fall_times)
print(f"Mean fall time = {mean_fall_time:.2f} seconds")
# Output: Mean fall time = 1.50 seconds
```

Code 5.98 Example of using NumPy in physics: create a simple array with data and compute the mean

Let's see the step-by-step explanation of this simple code. Once NumPy is installed, it is possible to import it into a Python script or interactive environment via `import numpy as np`. The acquired data are converted to an array by means of `np.array`, which is linked to the variable "fall_$ times". `np.mean` compute the arithmetic mean (average) of the given data stored as "fall_times". The estimated "mean_fall_time" is printed via the line #8 of the Code 5.98.

Following the idea on which we constructed the previous example, using NumPy, we can load data from a file representing the previous experiment and perform calculations:

```python
import numpy as np

# Load data from a file
loaded_data = np.load('data_file.npy')

# Perform calculations (replace this with your specific
    physics calculations)
mean_value = np.mean(loaded_data)
print(f"Mean value = {mean_value:.2f}")
# Output: Mean fall time = ...
```

Code 5.99 Example of using NumPy in physics by loading data from a file

The data stored in the file 'data_file.npy' are loaded with NumPy by means of the "load" or "loadtxt" functions—depending on the file extension—called by the command   np.load or   np.loadtxt , respectively.

5.5.1.1   Data Aggregation

NumPy has built-in aggregation functions for working with and computing summary statistics on arrays. The most common summary statistics in physics are the mean value and standard deviation, representing a dataset's "typical" values. Still, other aggregates (the sum, product, median, minimum, maximum, etc.) are also helpful. Aggregating along a single axis can usually be seen as reducing the dimensionality of our array by one. For instance, aggregating a one-dimensional array results in a single point, while aggregating a two-dimensional array yields a one-dimensional array. When dealing with multiple dimensions, we must choose which dimension (or axis) to eliminate. In NumPy, this is achieved by specifying the axis parameter in the function call. After constructing 1D, 2D and 3D arrays—named 'one_d', 'two_d' and 'three_d', respectively—, we will demonstrate the use of some Python NumPy aggregate functions, also providing a quick description of the functions themselves. In this frame, let's consider an example for each function inserting in the code few lines to show how aggregation leads to a reduction in the dimensionality of our starting array:

- np.sum() computes the sum of values of a given array. The reference to the elements of the array to be summed and other optional terms (e.g., axis or axes along which the sum is performed[4]) must be included inside the brackets.

```python
import numpy as np

# Create 1D, 2D and 3D arrays
print("\nGenerate 1D, 2D and 3D arrays")
one_d = np.array([1,2])
print("one_d =\n", one_d)
two_d = np.array([[3,4],[5,6]])
print("two_d =\n", two_d)
three_d = np.array([[[7,8],[9,10]],[[11,12],[13,14]]])
print("three_d =\n", three_d)

# Calculate the sum
print("\nCalculate the sum by means of the np.sum() function")
#Aggregating one dimensional arrays
sum1 = np.sum(one_d)
print("np.sum(one_d) =", sum1, ", with a dimension of", sum1.ndim)
```

---

[4] By default, setting axis=None will sum all elements in the input array. If the axis is negative, it will count from the last axis to the first.

```
17    # Output: np.sum(one_d) = 3 , with a dimension of 0
18    #Aggregating two dimensional arrays
19    sum2 = np.sum(two_d)
20    print("np.sum(two_d) =", sum2, ", with a dimension of",
      sum2.ndim)
21    # Output: np.sum(two_d) = 18 , with a dimension of 0
22    sum2_0 = np.sum(two_d, axis = 0)
23    print("np.sum(two_d, axis = 0) =", sum2_0, ", with a
      dimension of", sum2_0.ndim)
24    # Output: np.sum(two_d, axis = 0) = [ 8 10] , with a
      dimension of 1
25    sum2_1 = np.sum(two_d, axis = 1)
26    print("np.sum(two_d, axis = 1) =", sum2_1, ", with a
      dimension of", sum2_1.ndim)
27    # Output: np.sum(two_d, axis = 1) = [ 7 11] , with a
      dimension of 1
28    #Aggregating three dimensional arrays
29    sum3 = np.sum(three_d)
30    print("np.sum(three_d) =", sum3, ", with a dimension of"
      , sum3.ndim)
31    # Output: np.sum(three_d) = 84 , with a dimension of 0
32    sum3_0 = np.sum(three_d, axis = 0)
33    print("np.sum(three_d, axis = 0) =", sum3_0, ", with a
      dimension of", sum3_0.ndim)
34    # Output: np.sum(three_d, axis = 0) = [[18 20] [22 24]]
      , with a dimension of 2
35    sum3_1 = np.sum(three_d, axis = 1)
36    print("np.sum(three_d, axis = 1) =", sum3_1, ", with a
      dimension of", sum3_1.ndim)
37    # Output: np.sum(three_d, axis = 1) = [[16 18] [24 26]]
      , with a dimension of 2
38    sum3_2 = np.sum(three_d, axis = 2)
39    print("np.sum(three_d, axis = 2) =", sum3_2, ", with a
      dimension of", sum3_2.ndim)
40    # Output: np.sum(three_d, axis = 2) = [[15 19] [23 27]]
      , with a dimension of 2
41
42
```

Code 5.100 np.sum() example

In Example 5.100, it is shown how to apply the function to 1D, 2D and 3D arrays, distinguishing the results not only in those in which all elements add up, but also in those referring to specific axes. The output results of print() are also shown. It is evident that the dimensionality is lowered once the sum operation is performed, but it should be kept in mind that the dimensionality is reduced by 1 relative to the dimensionality of the starting array by indicating the axis along which to perform the sum. Note that write sum3_0 = np.sum(three_d, axis = 0) or sum3_0 = np.sum(three_d, 0) is the same thing.

- np.mean() returns the arithmetic mean of the array or an array with mean values along the specified axis. As seen above, if we do not indicate the axis on which to perform the operation, we will have as a result a scalar value.

```python
import numpy as np

# Create 1D, 2D and 3D arrays
print("\nGenerate 1D, 2D and 3D arrays")
one_d = np.array([1,2])
print("one_d =\n", one_d)
two_d = np.array([[3,4],[5,6]])
print("two_d =\n", two_d)
three_d = np.array([[[7,8],[9,10]],[[11,12],[13,14]]])
print("three_d =\n", three_d)

# Calculate the mean
print("\nCalculate the mean by means of the np.mean()
function")
#Aggregating one dimensional arrays
mean1 = np.mean(one_d)
print("np.mean(one_d) =", mean1, ", with a dimension of"
, mean1.ndim)
# Output: np.mean(one_d) = 1.5 , with a dimension of 0
#Aggregating two dimensional arrays
mean2 = np.mean(two_d)
print("np.mean(two_d) =", mean2, ", with a dimension of"
, mean2.ndim)
# Output: np.mean(two_d) = 4.5 , with a dimension of 0
mean2_0 = np.mean(two_d, axis = 0)
print("np.mean(two_d, axis = 0) =", mean2_0, ", with a
dimension of", mean2_0.ndim)
# Output: np.mean(two_d, axis = 0) = [4. 5.] , with a
dimension of 1
mean2_1 = np.mean(two_d, axis = 1)
print("np.mean(two_d, axis = 1) =", mean2_1, ", with a
dimension of", mean2_1.ndim)
# Output: np.mean(two_d, axis = 1) = [3.5 5.5] , with a
dimension of 1
#Aggregating three dimensional arrays
mean3 = np.mean(three_d)
print("np.mean(three_d) =", mean3, ", with a dimension
of", mean3.ndim)
# Output: np.mean(three_d) = 10.5 , with a dimension of
0
mean3_0 = np.mean(three_d, axis = 0)
print("np.mean(three_d, axis = 0) =", mean3_0, ", with a
 dimension of", mean3_0.ndim)
# Output: np.mean(three_d, axis = 0) = [[ 9. 10.] [11.
12.]] , with a dimension of 2
mean3_1 = np.mean(three_d, axis = 1)
print("np.mean(three_d, axis = 1) =", mean3_1, ", with a
 dimension of", mean3_1.ndim)
# Output: np.mean(three_d, axis = 1) = [[ 8.  9.] [12.
13.]] , with a dimension of 2
mean3_2 = np.mean(three_d, axis = 2)
print("np.mean(three_d, axis = 2) =", mean3_2, ", with a
 dimension of", mean3_2.ndim)
```

```
40   # Output: np.mean(three_d, axis = 2) = [[ 7.5   9.5]
     [11.5 13.5]] , with a dimension of 2

41
```

Code 5.101 np.mean() example

- np.average() calculates the weighted average of an array or along the specified axis. Compared with np.mean() , np.average() takes an optional weight parameter. If it is not supplied (i.e., if weights=None, all elements in the starting array are assumed to have a weight of one), these two functions are equivalent, providing the same results, as shown in the following example, which is to be compared with Example 5.101.

```
1    import numpy as np
2
3    # Create 1D, 2D and 3D arrays
4    print("\nGenerate 1D, 2D and 3D arrays")
5    one_d = np.array([1,2])
6    print("one_d =\n", one_d)
7    two_d = np.array([[3,4],[5,6]])
8    print("two_d =\n", two_d)
9    three_d = np.array([[[7,8],[9,10]],[[11,12],[13,14]]])
10   print("three_d =\n", three_d)
11
12   # Calculate the average
13   print("\nCalculate the average by means of the np.
     average() function")
14   #Aggregating one dimensional arrays
15   average1 = np.average(one_d)
16   print("np.average(one_d) =", average1, ", with a
     dimension of", average1.ndim)
17   # Output: np.average(one_d) = 1.5 , with a dimension of
     0
18   average1_w = np.average(one_d, weights=one_d)
19   print("np.average(one_d, weights=one_d) =", average1_w,
     ", with a dimension of", average1_w.ndim)
20   # Output: np.average(one_d, weights=one_d) =
     1.6666666666666667 , with a dimension of 0
21   #Aggregating two dimensional arrays
22   average2 = np.average(two_d)
23   print("np.average(two_d) =", average2, ", with a
     dimension of", average2.ndim)
24   # Output: np.average(two_d) = 4.5 , with a dimension of
     0
25   average2_w = np.average(two_d, weights=two_d)
26   print("np.average(two_d, weights=two_d) =", average2_w,
     ", with a dimension of", average2_w.ndim)
27   # Output: np.average(two_d, weights=two_d) =
     4.777777777777778 , with a dimension of 0
28   average2_0 = np.average(two_d, axis = 0)
```

```python
print("np.average(two_d, axis = 0) =", average2_0, ",
with a dimension of", average2_0.ndim)
# Output: np.average(two_d, axis = 0) = [4. 5.] , with a
 dimension of 1
average2_0_w = np.average(two_d, axis = 0, weights=two_d
)
print("np.average(two_d, axis = 0, weights=two_d) =",
average2_0_w, ", with a dimension of", average2_0_w.ndim
)
# Output: np.average(two_d, axis = 0, weights=two_d) =
[4.25 5.2 ] , with a dimension of 1
average2_1 = np.average(two_d, axis = 1)
print("np.average(two_d, axis = 1) =", average2_1, ",
with a dimension of", average2_1.ndim)
# Output: np.average(two_d, weights=two_d) =
4.777777777777778 , with a dimension of 0
average2_1_w = np.average(two_d, axis = 1, weights=two_d
)
print("np.average(two_d, axis = 1, weights=two_d) =",
average2_1_w, ", with a dimension of", average2_1_w.ndim
)
# Output: np.average(two_d, weights=two_d) =
4.777777777777778 , with a dimension of 0
#Aggregating three dimensional arrays
average3 = np.average(three_d)
print("np.average(three_d) =", average3, ", with a
dimension of", average3.ndim)
# Output:np.average(three_d) = 10.5 , with a dimension
of 0
average3_w = np.average(three_d, weights=three_d)
print("np.average(three_d, weights=three_d) =",
average3_w, ", with a dimension of", average3_w.ndim)
# Output: np.average(three_d, weights=three_d) = 11.0 ,
with a dimension of 0
average3_0 = np.average(three_d, axis = 0)
print("np.average(three_d, axis = 0) =", average3_0, ",
with a dimension of", average3_0.ndim)
# Output: np.average(three_d, axis = 0) = [[ 9. 10.]
[11. 12.]] , with a dimension of 2
average3_0_w = np.average(three_d, axis = 0, weights=
three_d)
print("np.average(three_d, axis = 0, weights=three_d) ="
, average3_0_w, ", with a dimension of", average3_0_w.
ndim)
# Output: np.average(three_d, axis = 0, weights=three_d)
 = [[ 9.44444444 10.4        ] [11.36363636 12.33333333]]
 , with a dimension of 2
average3_1 = np.average(three_d, axis = 1)
print("np.average(three_d, axis = 1) =", average3_1, ",
with a dimension of", average3_1.ndim)
# Output: np.average(three_d, axis = 1) = [[ 8.  9.]
[12. 13.]] , with a dimension of 2
```

```
56   average3_1_w = np.average(three_d, axis = 1, weights=
     three_d)
57   print("np.average(three_d, axis = 1, weights=three_d) ="
     , average3_1_w, ", with a dimension of", average3_1_w.
     ndim)
58   # Output: np.average(three_d, axis = 1, weights=three_d)
      = [[ 8.125        9.11111111] [12.08333333 13.07692308]]
      , with a dimension of 2
59   average3_2 = np.average(three_d, axis = 2)
60   print("np.average(three_d, axis = 2) =", average3_2, ",
     with a dimension of", average3_2.ndim)
61   # Output: np.average(three_d, axis = 2) = [[ 7.5  9.5]
     [11.5 13.5]] , with a dimension of 2
62   average3_2_w = np.average(three_d, axis = 2, weights=
     three_d)
63   print("np.average(three_d, axis = 2, weights=three_d) ="
     , average3_2_w, ", with a dimension of", average3_2_w.
     ndim)
64   # Output: np.average(three_d, axis = 2, weights=three_d)
      = [[ 7.53333333  9.52631579] [11.52173913 13.51851852]]
      , with a dimension of 2
65
```

Code 5.102   np.average() example

Values that result from weighted differ slightly from unweighted values and may
vary depending on the parameter because all elements may or may not have equal
weight. Note that the weights array must match the shape of the array containing
data to be averaged if no axis is specified (i.e., 1D, 2D or 3D array in our example).
The average is calculated as:

$$average = \frac{\sum(\text{elements} \cdot \text{weights})}{\sum \text{weights}} \tag{5.1}$$

and the only requirement for the weights is that their sum must not be zero.

- np.prod() returns the product of all elements in a given array. As seen above,
  even among the arguments of this function, we have the input array and the axis
  along which the product is calculated.

```
1    import numpy as np
2
3    # Create 1D, 2D and 3D arrays
4    print("\nGenerate 1D, 2D and 3D arrays")
5    one_d = np.array([1,2])
6    print("one_d =\n", one_d)
7    two_d = np.array([[3,4],[5,6]])
8    print("two_d =\n", two_d)
9    three_d = np.array([[[7,8],[9,10]],[[11,12],[13,14]]])
10   print("three_d =\n", three_d)
11
12   # Calculate the product
```

```
13    print("\nCalculate the product by means of the np.
      product() function")
14    #Aggregating one dimensional arrays
15    product1 = np.product(one_d)
16    print("np.product(one_d) =", product1, ", with a
      dimension of", product1.ndim)
17    # Output: np.product(one_d) = 2 , with a dimension of 0
18    #Aggregating two dimensional arrays
19    product2 = np.product(two_d)
20    print("np.product(two_d) =", product2, ", with a
      dimension of", product2.ndim)
21    # Output: np.product(two_d) = 360 , with a dimension of
      0
22    product2_0 = np.product(two_d, axis = 0)
23    print("np.product(two_d, axis = 0) =", product2_0, ",
      with a dimension of", product2_0.ndim)
24    # Output: np.product(two_d, axis = 0) = [15 24] , with a
       dimension of 1
25    product2_1 = np.product(two_d, axis = 1)
26    print("np.product(two_d, axis = 1) =", product2_1, ",
      with a dimension of", product2_1.ndim)
27    # Output: np.product(two_d, axis = 1) = [12 30] , with a
       dimension of 1
28    #Aggregating three dimensional arrays
29    product3 = np.product(three_d)
30    print("np.product(three_d) =", product3, ", with a
      dimension of", product3.ndim)
31    # Output: np.product(three_d) = 121080960 , with a
      dimension of 0
32    product3_0 = np.product(three_d, axis = 0)
33    print("np.product(three_d, axis = 0) =", product3_0, ",
      with a dimension of", product3_0.ndim)
34    # Output: np.product(three_d, axis = 0) = [[ 77  96]
      [117 140]] , with a dimension of 2
35    product3_1 = np.product(three_d, axis = 1)
36    print("np.product(three_d, axis = 1) =", product3_1, ",
      with a dimension of", product3_1.ndim)
37    # Output: np.product(three_d, axis = 1) = [[ 63  80]
      [143 168]] , with a dimension of 2
38    product3_2 = np.product(three_d, axis = 2)
39    print("np.product(three_d, axis = 2) =", product3_2, ",
      with a dimension of", product3_2.ndim)
40    # Output: np.product(three_d, axis = 2) = [[ 56  90]
      [132 182]] , with a dimension of 2
41
```

Code 5.103 `np.prod()` example

- `np.min()` and `np.max()` give the minimum and maximum values in an array or a given axis, respectively.

```python
import numpy as np

# Create 1D, 2D and 3D arrays
print("\nGenerate 1D, 2D and 3D arrays")
one_d = np.array([1,2])
print("one_d =\n", one_d)
two_d = np.array([[3,4],[5,6]])
print("two_d =\n", two_d)
three_d = np.array([[[7,8],[9,10]],[[11,12],[13,14]]])
print("three_d =\n", three_d)

# Calculate the min
print("\nCalculate the min value by means of the np.min
() function")
#Aggregating one dimensional arrays
min1 = np.min(one_d)
print("np.min(one_d) =", min1, ", with a dimension of",
min1.ndim)
# Output: np.min(one_d) = 1 , with a dimension of 0
#Aggregating two dimensional arrays
min2 = np.min(two_d)
print("np.min(two_d) =", min2, ", with a dimension of",
min2.ndim)
# Output: np.min(two_d) = 3 , with a dimension of 0
min2_0 = np.min(two_d, axis = 0)
print("np.min(two_d, axis = 0) =", min2_0, ", with a
dimension of", min2_0.ndim)
# Output: np.min(two_d, axis = 0) = [3 4] , with a
dimension of 1
min2_1 = np.min(two_d, axis = 1)
print("np.min(two_d, axis = 1) =", min2_1, ", with a
dimension of", min2_1.ndim)
# Output: np.min(two_d, axis = 1) = [3 5] , with a
dimension of 1
#Aggregating three dimensional arrays
min3 = np.min(three_d)
print("np.min(three_d) =", min3, ", with a dimension of"
, min3.ndim)
# Output: np.min(three_d) = 7 , with a dimension of 0
min3_0 = np.min(three_d, axis = 0)
print("np.min(three_d, axis = 0) =", min3_0, ", with a
dimension of", min3_0.ndim)
# Output: np.min(three_d, axis = 0) = [[ 7  8] [ 9 10]]
, with a dimension of 2
min3_1 = np.min(three_d, axis = 1)
print("np.min(three_d, axis = 1) =", min3_1, ", with a
dimension of", min3_1.ndim)
# np.min(three_d, axis = 1) = [[ 7  8] [11 12]] , with a
 dimension of 2
min3_2 = np.min(three_d, axis = 2)
print("np.min(three_d, axis = 2) =", min3_2, ", with a
dimension of", min3_2.ndim)
```

```python
40     # Output: np.min(three_d, axis = 2) = [[ 7  9] [11 13]]
       , with a dimension of 2
41
42     # Calculate the max
43     print("\nCalculate the max value by means of the np.max
       () function")
44     #Aggregating one dimensional arrays
45     max1 = np.max(one_d)
46     print("np.max(one_d) =", max1, ", with a dimension of",
       max1.ndim)
47     # Output: np.max(one_d) = 2 , with a dimension of 0
48     #Aggregating two dimensional arrays
49     max2 = np.max(two_d)
50     print("np.max(two_d) =", max2, ", with a dimension of",
       max2.ndim)
51     # Output: np.max(two_d) = 6 , with a dimension of 0
52     max2_0 = np.max(two_d, axis = 0)
53     print("np.max(two_d, axis = 0) =", max2_0, ", with a
       dimension of", max2_0.ndim)
54     # Output: np.max(two_d, axis = 0) = [5 6] , with a
       dimension of 1
55     max2_1 = np.max(two_d, axis = 1)
56     print("np.max(two_d, axis = 1) =", max2_1, ", with a
       dimension of", max2_1.ndim)
57     # Output: np.max(two_d, axis = 1) = [4 6] , with a
       dimension of 1
58     #Aggregating three dimensional arrays
59     max3 = np.max(three_d)
60     print("np.max(three_d) =", max3, ", with a dimension of"
       , max3.ndim)
61     # Output: np.max(three_d) = 14 , with a dimension of 0
62     max3_0 = np.max(three_d, axis = 0)
63     print("np.max(three_d, axis = 0) =", max3_0, ", with a
       dimension of", max3_0.ndim)
64     # Output: np.max(three_d, axis = 0) = [[11 12] [13 14]]
       , with a dimension of 2
65     max3_1 = np.max(three_d, axis = 1)
66     print("np.max(three_d, axis = 1) =", max3_1, ", with a
       dimension of", max3_1.ndim)
67     # Output: np.max(three_d, axis = 1) = [[ 9 10] [13 14]]
       , with a dimension of 2
68     max3_2 = np.max(three_d, axis = 2)
69     print("np.max(three_d, axis = 2) =", max3_2, ", with a
       dimension of", max3_2.ndim)
70     # Output: np.max(three_d, axis = 2) = [[ 8 10] [12 14]]
       , with a dimension of 2
71
```

Code 5.104   np.min() and np.max() example

- np.minimum() and np.maximum() cross-check two arrays' elements in a

one-to-one comparison to construct an array of minimum and maximum values,
respectively. We need two mandatory parameters referring to two input arrays to

be compared, and then we have the chance to insert several optional parameters. The result of np.minimum() (or np.maximum()) is a new array containing the element-wise minima (or the maxima), furnishing a scalar if the elements of the input arrays are scalars.

```python
import numpy as np

# Calculate element-wise minimum of array elements
print("\nCalculate the minimum elements between two
    arrays by means of the np.minimum() function")
minimum = np.minimum([2, 3, 4], [1, 5, 2])
print("np.minimum([2, 3, 4], [1, 5, 2]) =", minimum, ",
    with a dimension of", minimum.ndim)
# Output: np.minimum([2, 3, 4], [1, 5, 2]) = [1 3 2] ,
    with a dimension of 1

# Calculate element-wise maximum of array elements
print("\nCalculate the maximum elements between two
    arrays by means of the np.maximum() function")
maximum = np.maximum([2, 3, 4], [1, 5, 2])
print("np.maximum([2, 3, 4], [1, 5, 2]) =", maximum, ",
    with a dimension of", maximum.ndim)
# Output: np.maximum([2, 3, 4], [1, 5, 2]) = [2 5 4] ,
    with a dimension of 1

```

Code 5.105 np.minimum() and np.maximum() example

- np.median() returns the median[5] of the array elements.

```python
import numpy as np

# Create 1D, 2D and 3D arrays
print("\nGenerate 1D, 2D and 3D arrays")
one_d = np.array([11,8,19,1])
print("one_d =\n", one_d)
two_d = np.array([[6,1,11,8],[7,6,2,15]])
print("two_d =\n", two_d)
three_d = np.array
    ([[[7,13,8],[9,1,10]],[[11,12,2],[2,13,14]]])
print("three_d =\n", three_d)

# Calculate the median
print("\nCalculate the median by means of the np.median
    () function")
#Aggregating one dimensional arrays
median1 = np.median(one_d)
```

---

[5] In statistics, it is a position index indicating the term placed in the middle of the distribution. To calculate the median of a set of data points, arrange them in ascending order. If the total number of terms is odd, the median is the middle term. If the total number of terms is even, the median is the average of the two middle terms.

```
16    print("np.median(one_d) =", median1, ", with a dimension
         of", median1.ndim)
17    # Output: np.median(one_d) = 9.5 , with a dimension of 0
18    #Aggregating two dimensional arrays
19    median2 = np.median(two_d)
20    print("np.median(two_d) =", median2, ", with a dimension
         of", median2.ndim)
21    # Output: np.median(two_d) = 6.5 , with a dimension of 0
22    median2_0 = np.median(two_d, axis = 0)
23    print("np.median(two_d, axis = 0) =", median2_0, ", with
         a dimension of", median2_0.ndim)
24    # Output: np.median(two_d, axis = 0) = [ 6.5   3.5   6.5
         11.5] , with a dimension of 1
25    median2_1 = np.median(two_d, axis = 1)
26    print("np.median(two_d, axis = 1) =", median2_1, ", with
         a dimension of", median2_1.ndim)
27    # Output: np.median(two_d, axis = 1) = [7.   6.5] , with
         a dimension of 1
28    #Aggregating three dimensional arrays
29    median3 = np.median(three_d)
30    print("np.median(three_d) =", median3, ", with a
         dimension of", median3.ndim)
31    # Output: np.median(three_d) = 9.5 , with a dimension of
          0
32    median3_0 = np.median(three_d, axis = 0)
33    print("np.median(three_d, axis = 0) =", median3_0, ",
         with a dimension of", median3_0.ndim)
34    # Output: np.median(three_d, axis = 0) = [[ 9.   12.5   5.
         ] [ 5.5   7.   12. ]] , with a dimension of 2
35    median3_1 = np.median(three_d, axis = 1)
36    print("np.median(three_d, axis = 1) =", median3_1, ",
         with a dimension of", median3_1.ndim)
37    # np.median(three_d, axis = 1) = [[ 8.   7.   9. ] [ 6.5
         12.5  8. ]] , with a dimension of 2
38    median3_2 = np.median(three_d, axis = 2)
39    print("np.median(three_d, axis = 2) =", median3_2, ",
         with a dimension of", median3_2.ndim)
40    # Output: np.median(three_d, axis = 2) = [[ 8.   9.] [11.
         13.]] , with a dimension of 2
41
```

Code 5.106   np.median() example

- np.var() furnishes the variance[6] of an array or axis.

```
1    import numpy as np
2
3    # Create 1D, 2D and 3D arrays
4    print("\nGenerate 1D, 2D and 3D arrays")
5    one_d = np.array([1,2])
6    print("one_d =\n", one_d)
```

---

[6] Variance is the expected value of the squared deviation from the mean of a random variable, measuring how far a set of numbers is spread out from their average value (dispersion).

```python
two_d = np.array([[3,4],[5,6]])
print("two_d =\n", two_d)
three_d = np.array([[[7,8],[9,10]],[[11,12],[13,14]]])
print("three_d =\n", three_d)


# Calculate the variance
print("\nCalculate the variance by means of the np.var()
  function")
#Aggregating one dimensional arrays
var1 = np.var(one_d)
print("np.var(one_d) =", var1, ", with a dimension of",
var1.ndim)
# Output: np.var(one_d) = 0.25 , with a dimension of 0
#Aggregating two dimensional arrays
var2 = np.var(two_d)
print("np.var(two_d) =", var2, ", with a dimension of",
var2.ndim)
# Output: np.var(two_d) = 1.25 , with a dimension of 0
var2_0 = np.var(two_d, axis = 0)
print("np.var(two_d, axis = 0) =", var2_0, ", with a
dimension of", var2_0.ndim)
# Output: np.var(two_d, axis = 0) = [1. 1.] , with a
dimension of 1
var2_1 = np.var(two_d, axis = 1)
print("np.var(two_d, axis = 1) =", var2_1, ", with a
dimension of", var2_1.ndim)
# Output: np.var(two_d, axis = 1) = [0.25 0.25] , with a
 dimension of 1
#Aggregating three dimensional arrays
var3 = np.var(three_d)
print("np.var(three_d) =", var3, ", with a dimension of"
, var3.ndim)
# Output: np.var(three_d) = 5.25 , with a dimension of 0
var3_0 = np.var(three_d, axis = 0)
print("np.var(three_d, axis = 0) =", var3_0, ", with a
dimension of", var3_0.ndim)
# Output: np.var(three_d, axis = 0) = [[4. 4.] [4. 4.]]
, with a dimension of 2
var3_1 = np.var(three_d, axis = 1)
print("np.var(three_d, axis = 1) =", var3_1, ", with a
dimension of", var3_1.ndim)
# Output: np.var(three_d, axis = 1) = [[1. 1.] [1. 1.]]
, with a dimension of 2
var3_2 = np.var(three_d, axis = 2)
print("np.var(three_d, axis = 2) =", var3_2, ", with a
dimension of", var3_2.ndim)
# Output: np.var(three_d, axis = 2) = [[0.25 0.25] [0.25
 0.25]] , with a dimension of 2

```

Code 5.107   np.var() example

- np.std() computes the standard deviation[7]

```python
import numpy as np

# Create 1D, 2D and 3D arrays
print("\nGenerate 1D, 2D and 3D arrays")
one_d = np.array([1,2])
print("one_d =\n", one_d)
two_d = np.array([[3,4],[5,6]])
print("two_d =\n", two_d)
three_d = np.array([[[7,8],[9,10]],[[11,12],[13,14]]])
print("three_d =\n", three_d)

# Calculate the std
print("\nCalculate the std by means of the np.std() function")
#Aggregating one dimensional arrays
std1 = np.std(one_d)
print("np.std(one_d) =", std1, ", with a dimension of", std1.ndim)
# Output: np.std(one_d) = 0.5 , with a dimension of 0
#Aggregating two dimensional arrays
std2 = np.std(two_d)
print("np.std(two_d) =", std2, ", with a dimension of", std2.ndim)
# Output: np.std(two_d) = 1.118033988749895 , with a dimension of 0
std2_0 = np.std(two_d, axis = 0)
print("np.std(two_d, axis = 0) =", std2_0, ", with a dimension of", std2_0.ndim)
# Output: np.std(two_d, axis = 0) = [1. 1.] , with a dimension of 1
std2_1 = np.std(two_d, axis = 1)
print("np.std(two_d, axis = 1) =", std2_1, ", with a dimension of", std2_1.ndim)
# Output: np.std(two_d, axis = 1) = [0.5 0.5] , with a dimension of 1
#Aggregating three dimensional arrays
std3 = np.std(three_d)
print("np.std(three_d) =", std3, ", with a dimension of", std3.ndim)
# Output: np.std(three_d) = 2.29128784747792 , with a dimension of 0
std3_0 = np.std(three_d, axis = 0)
print("np.std(three_d, axis = 0) =", std3_0, ", with a dimension of", std3_0.ndim)
# Output: np.std(three_d, axis = 0) = [[2. 2.] [2. 2.]] , with a dimension of 2
std3_1 = np.std(three_d, axis = 1)
print("np.std(three_d, axis = 1) =", std3_1, ", with a dimension of", std3_1.ndim)
```

---

[7] Standard Deviation or mean-square deviation is obtained as the square root of the variance, and it measures the spread of data distribution in the given data set. In physics, this parameter indicates the random error of measuring a physical quantity in a given array or along a given axis.

```
37   # Output: np.std(three_d, axis = 1) = [[1. 1.] [1. 1.]]
     , with a dimension of 2
38   std3_2 = np.std(three_d, axis = 2)
39   print("np.std(three_d, axis = 2) =", std3_2, ", with a
     dimension of", std3_2.ndim)
40   # Output: np.std(three_d, axis = 2) = [[0.5 0.5] [0.5
     0.5]] , with a dimension of 2
41
```

Code 5.108   np.std() example

- np.cumsum() and np.cumprod() result in the cumulative sum and product of a given array or in a given axis, respectively.

```
1    import numpy as np
2
3    # Create 1D, 2D and 3D arrays
4    print("\nGenerate 1D, 2D and 3D arrays")
5    one_d = np.array([1,2])
6    print("one_d =\n", one_d)
7    two_d = np.array([[3,4],[5,6]])
8    print("two_d =\n", two_d)
9    three_d = np.array([[[7,8],[9,10]],[[11,12],[13,14]]])
10   print("three_d =\n", three_d)
11
12   # Calculate the cumulative sum
13   print("\nCalculate the cumulative sum by means of the np
     .cumsum() function")
14   #Aggregating one dimensional arrays
15   cumsum1 = np.cumsum(one_d)
16   print("np.cumsum(one_d) =", cumsum1, ", with a dimension
     of", cumsum1.ndim)
17   # Output: np.cumsum(one_d) = [1 3] , with a dimension of
     1
18   #Aggregating two dimensional arrays
19   cumsum2 = np.cumsum(two_d)
20   print("np.cumsum(two_d) =", cumsum2, ", with a dimension
     of", cumsum2.ndim)
21   # Output: np.cumsum(two_d) = [ 3  7 12 18] , with a
     dimension of 1
22   cumsum2_0 = np.cumsum(two_d, axis = 0)
23   print("np.cumsum(two_d, axis = 0) =", cumsum2_0, ", with
     a dimension of", cumsum2_0.ndim)
24   # Output: np.cumsum(two_d, axis = 0) = [[ 3  4] [ 8 10]]
     , with a dimension of 2
25   cumsum2_1 = np.cumsum(two_d, axis = 1)
26   print("np.cumsum(two_d, axis = 1) =", cumsum2_1, ", with
     a dimension of", cumsum2_1.ndim)
27   # Output: np.cumsum(two_d, axis = 1) = [[ 3  7] [ 5 11]]
     , with a dimension of 2
28   #Aggregating three dimensional arrays
29   cumsum3 = np.cumsum(three_d)
30   print("np.cumsum(three_d) =", cumsum3, ", with a
     dimension of", cumsum3.ndim)
```

```python
31  # Output: np.cumsum(three_d) = [ 7 15 24 34 45 57 70 84]
    , with a dimension of 1
32  cumsum3_0 = np.cumsum(three_d, axis = 0)
33  print("np.cumsum(three_d, axis = 0) =", cumsum3_0, ",
    with a dimension of", cumsum3_0.ndim)
34  # Output: np.cumsum(three_d, axis = 0) = [[[ 7  8] [ 9
    10]] [[18 20] [22 24]]] , with a dimension of 3
35  cumsum3_1 = np.cumsum(three_d, axis = 1)
36  print("np.cumsum(three_d, axis = 1) =", cumsum3_1, ",
    with a dimension of", cumsum3_1.ndim)
37  # Output: np.cumsum(three_d, axis = 1) = [[[ 7  8] [16
    18]] [[11 12] [24 26]]] , with a dimension of 3
38  cumsum3_2 = np.cumsum(three_d, axis = 2)
39  print("np.cumsum(three_d, axis = 2) =", cumsum3_2, ",
    with a dimension of", cumsum3_2.ndim)
40  # Output: np.cumsum(three_d, axis = 2) = [[[ 7 15] [ 9
    19]] [[11 23] [13 27]]] , with a dimension of 3
41
42  # Calculate the cumulative product
43  print("\nCalculate the cumulative product by means of
    the np.cumprod() function")
44  #Aggregating one dimensional arrays
45  cumprod1 = np.cumprod(one_d)
46  print("np.cumprod(one_d) =", cumprod1, ", with a
    dimension of", cumprod1.ndim)
47  # Output: np.cumprod(one_d) = [1 2] , with a dimension
    of 1
48  #Aggregating two dimensional arrays
49  cumprod2 = np.cumprod(two_d)
50  print("np.cumprod(two_d) =", cumprod2, ", with a
    dimension of", cumprod2.ndim)
51  # Output: np.cumprod(two_d) = [  3  12  60 360] , with a
     dimension of 1
52  cumprod2_0 = np.cumprod(two_d, axis = 0)
53  print("np.cumprod(two_d, axis = 0) =", cumprod2_0, ",
    with a dimension of", cumprod2_0.ndim)
54  # Output: np.cumprod(two_d, axis = 0) = [[ 3  4] [15
    24]] , with a dimension of 2
55  cumprod2_1 = np.cumprod(two_d, axis = 1)
56  print("np.cumprod(two_d, axis = 1) =", cumprod2_1, ",
    with a dimension of", cumprod2_1.ndim)
57  # Output: np.cumprod(two_d, axis = 1) = [[ 3 12] [ 5
    30]] , with a dimension of 2
58  #Aggregating three dimensional arrays
59  cumprod3 = np.cumprod(three_d)
60  print("np.cumprod(three_d) =", cumprod3, ", with a
    dimension of", cumprod3.ndim)
61  # Output: np.cumprod(three_d) = [        7        56
       504      5040     55440    665280   8648640
    121080960] , with a dimension of 1
62  cumprod3_0 = np.cumprod(three_d, axis = 0)
63  print("np.cumprod(three_d, axis = 0) =", cumprod3_0, ",
    with a dimension of", cumprod3_0.ndim)
```

```
64    # Output: np.cumprod(three_d, axis = 0) = [[[  7    8] [
       9  10]] [[ 77   96] [117 140]]] , with a dimension of 3
65    cumprod3_1 = np.cumprod(three_d, axis = 1)
66    print("np.cumprod(three_d, axis = 1) =", cumprod3_1, ",
       with a dimension of", cumprod3_1.ndim)
67    # Output: np.cumprod(three_d, axis = 1) = [[[  7    8] [
       63  80]] [[ 11   12] [143 168]]] , with a dimension of 3
68    cumprod3_2 = np.cumprod(three_d, axis = 2)
69    print("np.cumprod(three_d, axis = 2) =", cumprod3_2, ",
       with a dimension of", cumprod3_2.ndim)
70    # Output: np.cumprod(three_d, axis = 2) = [[[  7   56] [
       9  90]] [[ 11 132] [ 13 182]]] , with a dimension of 3
71
```

Code 5.109 `np.cumsum()` and `np.cumprod()` example

- `np.percentile()` returns the percentile[8] (based on the given value) of an array or an axis. In this case, the parameters required by the function are (i) data placed in input arrays or an object to be converted to an array, and (ii) a value q from 0 to 100 representing a percentage or a subsequence of percentages to be computed by percentile. Note that `np.percentile()` returns the q-th percentile(s) of the array elements.

```
1     import numpy as np
2
3     # Create 1D, 2D and 3D arrays
4     print("\nGenerate 1D, 2D and 3D arrays")
5     one_d = np.array([11,8,19,1])
6     print("one_d =\n", one_d)
7     two_d = np.array([[6,1,11,8],[7,6,2,15]])
8     print("two_d =\n", two_d)
9     three_d = np.array
       ([[[7,13,8],[9,1,10]],[[11,12,2],[2,13,14]]])
10    print("three_d =\n", three_d)
11
12    # Calculate the percentile
13    print("\nCalculate the percentile by means of the np.
       percentile() function")
14    #Aggregating one dimensional arrays
15    percentile1_0 = np.percentile(one_d, 0)
16    print("np.percentile(one_d, 0) =", percentile1_0, ",
       with a dimension of", percentile1_0.ndim)
17    # Output: np.percentile(one_d, 0) = 1.0 , with a
       dimension of 0
18    percentile1_50 = np.percentile(one_d, 50)
19    print("np.percentile(one_d, 50) =", percentile1_50, ",
       with a dimension of", percentile1_50.ndim)
20    # Output: np.percentile(one_d, 50) = 9.5 , with a
       dimension of 0
21    #Aggregating two dimensional arrays
```

---

[8] In statistics, a measure used to indicate the minimum value below which a given percentage of the other elements under observation falls.

```python
percentile2_0 = np.percentile(two_d, 0)
print("np.percentile(two_d, 0) =", percentile2_0, ",
    with a dimension of", percentile2_0.ndim)
# Output: np.percentile(two_d, 0) = 1.0 , with a
    dimension of 0
percentile2_50 = np.percentile(two_d, 50)
print("np.percentile(two_d, 50) =", percentile2_50, ",
    with a dimension of", percentile2_50.ndim)
# Output: np.percentile(two_d, 50) = 6.5 , with a
    dimension of 0
percentile2_0_0 = np.percentile(two_d, 0, axis = 0)
print("np.percentile(two_d, 0, axis = 0) =",
    percentile2_0_0, ", with a dimension of",
    percentile2_0_0.ndim)
# Output: np.percentile(two_d, 0, axis = 0) = [6. 1. 2.
    8.] , with a dimension of 1
percentile2_50_0 = np.percentile(two_d, 50, axis = 0)
print("np.percentile(two_d, 50, axis = 0) =",
    percentile2_50_0, ", with a dimension of",
    percentile2_50_0.ndim)
# Output: np.percentile(two_d, 50, axis = 0) = [ 6.5
    3.5  6.5 11.5] , with a dimension of 1
percentile2_0_1 = np.percentile(two_d, 0, axis = 1)
print("np.percentile(two_d, 0, axis = 1) =",
    percentile2_0_1, ", with a dimension of",
    percentile2_0_1.ndim)
# Output: np.percentile(two_d, 0, axis = 1) = [1. 2.] ,
    with a dimension of 1
percentile2_50_1 = np.percentile(two_d, 50, axis = 1)
print("np.percentile(two_d, 50, axis = 1) =",
    percentile2_50_1, ", with a dimension of",
    percentile2_50_1.ndim)
# Output: np.percentile(two_d, 50, axis = 1) = [7.   6.5]
    , with a dimension of 1
#Aggregating three dimensional arrays
percentile3_0 = np.percentile(three_d, 0)
print("np.percentile(three_d, 0) =", percentile3_0, ",
    with a dimension of", percentile3_0.ndim)
# Output: np.percentile(three_d, 0) = 1.0 , with a
    dimension of 0
percentile3_50 = np.percentile(three_d, 50)
print("np.percentile(three_d, 50) =", percentile3_50, ",
    with a dimension of", percentile3_50.ndim)
# Output: np.percentile(three_d, 50) = 9.5 , with a
    dimension of 0
percentile3_0_0 = np.percentile(three_d, 0, axis = 0)
print("np.percentile(three_d, 0, axis = 0) =",
    percentile3_0_0, ", with a dimension of",
    percentile3_0_0.ndim)
# Output: np.percentile(three_d, 0, axis = 0) = [[ 7.
    12.   2.] [ 2.   1.  10.]] , with a dimension of 2
percentile3_50_0 = np.percentile(three_d, 50, axis = 0)
```

```
51   print("np.percentile(three_d, 50, axis = 0) =",
     percentile3_50_0, ", with a dimension of",
     percentile3_50_0.ndim)
52   # Output: np.percentile(three_d, 50, axis = 0) = [[ 9.
     12.5   5. ] [ 5.5   7.   12. ]] , with a dimension of 2
53   percentile3_0_1 = np.percentile(three_d, 0, axis = 1)
54   print("np.percentile(three_d, 0, axis = 1) =",
     percentile3_0_1, ", with a dimension of",
     percentile3_0_1.ndim)
55   # Output: np.percentile(three_d, 0, axis = 1) = [[ 7.
     1.   8.] [ 2.  12.   2.]] , with a dimension of 2
56   percentile3_50_1 = np.percentile(three_d, 50, axis = 1)
57   print("np.percentile(three_d, 50, axis = 1) =",
     percentile3_50_1, ", with a dimension of",
     percentile3_50_1.ndim)
58   # Output: np.percentile(three_d, 50, axis = 1) = [[ 8.
      7.    9. ] [ 6.5 12.5   8. ]] , with a dimension of 2
59   percentile3_0_2 = np.percentile(three_d, 0, axis = 2)
60   print("np.percentile(three_d, 0, axis = 2) =",
     percentile3_0_2, ", with a dimension of",
     percentile3_0_2.ndim)
61   # Output: np.percentile(three_d, 0, axis = 2) = [[7. 1.]
      [2. 2.]] , with a dimension of 2
62   percentile3_50_2 = np.percentile(three_d, 50, axis = 2)
63   print("np.percentile(three_d, 50, axis = 2) =",
     percentile3_50_2, ", with a dimension of",
     percentile3_50_2.ndim)
64   # Output: np.percentile(three_d, 50, axis = 2) = [[ 8.
     9.] [11. 13.]] , with a dimension of 2
65
```

Code 5.110 `np.percentile()` example

5.5.1.2 Data Cleaning

Within the data manipulation, NumPy allows for careful cleaning of the data with well-defined functions. In fact, working with data, it could be really useful to ensure that your data does not contain NaN or infinity values, and to replace them with specific values. Examples of such functions are given by:

- `isnan()` is used to check for NaN (Not a Number) values in an array.

- `nan_to_num()` is employed to change NaN (Not a Number) and infinity values with zero and finite values, respectively.

- `percentile()` determines the value below which a specified percentage of observations in a dataset fall. This statistical function is frequently used to identify and manage outliers by setting a threshold beyond which data points are deemed extreme.

Here's an example of how to use   numpy.nan_to_num() :

```python
import numpy as np

# Create an array with NaN, positive infinity, and negative
    infinity
data_to_clean = np.array([np.nan, 1, 2, np.nan, np.inf, -np.
    inf, 3, 4, np.inf, -np.inf])

# Use nan_to_num to replace NaN with 0, positive infinity with
    a large number, and negative infinity with a small number
cleaned_data = np.nan_to_num(data_to_clean, nan=0.0, posinf=1
    e6, neginf=-1e6)

print("Original Data:")
print(data_to_clean)
# Output: Original Data:
#           [ nan   1.   2.   nan  inf -inf   3.   4.   inf -inf]

print("\nCleaned Data:")
print(cleaned_data)
# Output: Cleaned Data:
#           [ 0.e+00  1.e+00  2.e+00  0.e+00  1.e+06 -1.e+06
#             3.e+00  4.e+00  1.e+06  -1.e+06]
```

Code 5.111 Example of using NumPy for Data Cleaning

In this example,   np.nan_to_num()   replaces "NaN" values with 0.0, positive
infinity (np.inf) values with 1×10^6, and negative infinity (-np.inf) values with
-1×10^6.

5.5.1.3   Data Transformation

NumPy includes methods to perform "Data transformation", a series of pre-processing techniques for transforming raw data. *Standardization*, *scaling* and *normalization* are some processes used to rescale data.

Standardization is the technique that allows rescaling data with zero mean and unit variance. In NumPy, this may be attained by using the combination of two fundamental functions in statistics, i.e.,   np.mean() —already seen in Example 5.98—and   np.std()   that computes the standard deviation of the given data (array elements).

```python
import numpy as np

# Sample dataset: temperature readings from different sensors
    (in Celsius)
temperatures = np.array([14.8, 15.2, 16.8, 14.5, 17.1, 15.9,
    16.3, 17.0, 15.4, 16.1])
```

```python
5
6  # Calculate the mean and standard deviation
7  mean = np.mean(temperatures)
8  std_dev = np.std(temperatures)
9
10 # Standardize the dataset
11 standardized_temperatures = (temperatures - mean) / std_dev
12
13 print("Original Data (Celsius):")
14 print(temperatures)
15 # Output: Original Data (Celsius):
16 #         [14.8 15.2 16.8 14.5 17.1 15.9 16.3 17.  15.4 16.1]
17
18 print("\nStandardized Data:")
19 print(standardized_temperatures)
20 # Output: Standardized Data:
21 #         [-1.27586207 -0.81609195  1.02298851 -1.62068966
       1.36781609
22 #          -0.01149425   0.44827586  1.25287356 -0.5862069
       0.2183908 ]
23
24 # Checking the validity of the standardization procedure
       carried out
25 print("Mean of standardized temperatures:")
26 print(np.mean(standardized_temperatures))
27 # Output: Mean of standardized temperatures: 4.468647674116255
       e-16
28
29 print("\nStandard deviation of standardized temperatures:")
30 print(np.std(standardized_temperatures))
31 # Output: Standard deviation of standardized temperatures: 1.0
```

Code 5.112 Standardization of data with NumPy

Example 5.112 calculates the mean and standard deviation of an array of temperature readings. After that, we standardize the data by subtracting the mean and dividing by the standard deviation for each reading. This process transforms the data so that it has a mean of 0 and a standard deviation of 1—as shown in the final part of the code, "Checking the validity of the standardization procedure carried out." This makes it easier to compare and analyze the standardized data with other similar experiments.

Scaling is a method used to adjust data to fit within a specific range. Usually, we may use:

- the *maximum absolute scaling* process that rescales each element between −1 and 1 by dividing every value by its maximum absolute value;
- the *min-max feature scaling* procedure—also known as *normalization*—in which we rescale the feature to a range of [0, 1].

Example 5.113 shows how to apply the maximum absolute scaling procedure in NumPy using the combination of `.max()` and `.abs()` methods. This latter function computes the non-negative value element-wise of an input array, furnishing an ndarray containing the absolute value of each element as a result.

```python
import numpy as np

# Sample dataset: modules of the acceleration vector
acceleration = np.array([2.5, 3.6, -4.5, -1.2, 1.1, 2.9, 1.7,
    2.0, 3.4, -1.5])

# Calculate the absolute and maximum values
abs_val = np.abs(acceleration)
max_abs_val = np.max(abs_val)

# Scale the dataset to the range [-1, 1]
scaled_acceleration = acceleration / max_abs_val

print("Original Data:")
print(acceleration)
# Output: Original Data:
#           [ 2.5  3.6 -4.5 -1.2  1.1  2.9  1.7  2.   3.4 -1.5]

print("\nScaled Data [-1, 1]:")
print(scaled_acceleration)
# Output: Scaled Data [-1, 1]:
#           [ 0.55555556  0.8         -1.          -0.26666667
#             0.24444444  0.64444444  0.37777778  0.44444444
#             0.75555556  -0.33333333]
```

Code 5.113 Scaling of data with the maximum absolute scaling procedure in NumPy

As for the normalization procedure, it is shown following two different methodologies. In the first method named "the min-max feature scaling", we make use of np.min() and np.max() functions scaling the data as shown in the Example 5.114.

```python
import numpy as np

# Sample dataset: voltage readings from different sensors (in
    volts)
voltage = np.array([2.5, 3.6, 1.8, 4.5, 3.1, 2.9, 1.7, 2.0,
    3.4, 1.5])

# Calculate the minimum and maximum values
min_val = np.min(voltage)
max_val = np.max(voltage)

# Scale the dataset to the range [0, 1]
scaled_voltage = (voltage - min_val) / (max_val - min_val)

print("Original Data (Volts):")
print(voltage)
# Output: Original Data (Volts):
#           [2.5 3.6 1.8 4.5 3.1 2.9 1.7 2.  3.4 1.5]
```

```
18  print("\nScaled Data [0, 1]:")
19  print(scaled_voltage)
20  # Output: Scaling Data:
21  #          [0.33333333 0.7          0.1          1.
        0.53333333
22  #           0.46666667  0.06666667 0.16666667 0.63333333 0.
            ]
```

Code 5.114 Normalization of data with the min-max feature scaling in NumPy

Another method to perform *normalization* establishes the rescaling of the data so that it has a unit norm. In the following example, we will use the   np.linalg.norm() function to computes the L2 norm[9] of a set of data that consists of an array of force readings. The normalized data is then obtained by dividing each reading by its L2 norm, resulting in a dataset where the sum of the squares of the values equals.

```
1  import numpy as np
2
3  # Sample dataset: voltage readings from different sensors (in
       volts)
4  voltage = np.array([2.5, 3.6, 1.8, 4.5, 3.1, 2.9, 1.7, 2.0,
       3.4, 1.5])
5
6  # Calculate the L2 norm of the dataset
7  l2_norm = np.linalg.norm(voltage)
8
9  # Normalize the dataset
10 normalized_data = voltage / l2_norm
11
12 print("Original Data (Volts):")
13 print(voltage)
14 # Output: Original Data (Volts):
15 #          [2.5 3.6 1.8 4.5 3.1 2.9 1.7 2.  3.4 1.5]
16
17 print("\nL2 norm:")
18 print(l2_norm)
19 # Output: L2 norm:
20 #          9.023303164584464
21
22 print("\nNormalized Data:")
23 print(normalized_data)
24 # Output: Normalized Data:
25 #          [0.2770604  0.39896698 0.19948349 0.49870872
26 #           0.3435549  0.32139007 0.18840107 0.22164832
27 #           0.37680215 0.16623624]
```

Code 5.115 Normalization of data with the np.linalg.norm() function in NumPy

As shown in the outputs of the two examples immediately above, the normalized data fall within the range [0, 1], ensuring comparability and preventing any variable from dominating the analysis due to its magnitude.

---

[9] Also known as the "Euclidean norm", it is the shortest distance to go from one point to another calculated as the square root of the sum of the squared magnitudes of the vectors in a space.

5.5.1.4   Data Filtering

The procedure used to manually select some elements out of an existing array and then create a new array out of them is called *filtering*. We have several ways to obtain filtered data by using NumPy. Now, we will show a simple code that contains it examples of filtering temporal data from a physics experiment.

```python
import numpy as np

# Sample dataset: time readings from several stopwatches (in
    seconds)
recorded_times = np.array([5.0, 3.1, 7.2, 3.3, 8.4, 2.5, 4.2,
    3.8, 9.1, 3.9])
print("Time readings:")
print(recorded_times)
# Output: Time readings:
#          [5.   3.1 7.2 3.3 8.4 2.5 4.2 3.8 9.1 3.9]

# From "recorded_times" we want to select only times greater
    than 4 seconds
print('\nFrom "Time readings", we want to select only times
    greater than 4 seconds.')
# Output: From "Time readings", we want to select only times
    greater than 4 seconds.

# Method 1. Filter values by constructing a list of indices to
    use as a selector
indices = [0, 2, 4, 6, 8]
method1 = recorded_times[indices]
print("\nFiltering with method 1 (indices):")
print(method1)
# Output: Filtering with method 1 (indices):
#          [5.   7.2 8.4 4.2 9.1]

# Method 2. Filter values by constructing a list of True and
    False (boolean array)
boolean_array = [True, False, True, False, True, False, True,
    False, True, False]
method2 = recorded_times[boolean_array]
print("\nFiltering with method 2 (boolean array):")
print(method2)
# Output: Filtering with method 2 (boolean array):
#          [5.   7.2 8.4 4.2 9.1]

# Method 3. Filter array based on conditions
filtered_arr = []
# go through each element in arr
for element in recorded_times:
    # if the element is higher than 4, set the value to True,
    otherwise False:
    if element > 4:
        filtered_arr.append(True)
    else:
```

```
38          filtered_arr.append(False)
39 method3 = recorded_times[filtered_arr]
40 print("\nFiltering with method 3 (filter array with conditions
       ):")
41 print(method3)
42 # Output: Filtering with method 3 (filter array with
       conditions):
43 #          [5.  7.2 8.4 4.2 9.1]
44
45 # Method 4. Filter array with iterable variable in our
       condition
46 mask = recorded_times > 4
47 method4 = recorded_times[mask]
48 print("\nFiltering with method 4 (filter array with iterable
       variable):")
49 print(method4)
50 # Output: Filtering with method 4 (filter array with iterable
       variable):
51 #          [5.  7.2 8.4 4.2 9.1]
```

Code 5.116 Data filtering of data with NumPy

5.5.1.5 Summary

In summary, NumPy empowers Python users with powerful array manipulation capabilities, making it a cornerstone of scientific computing.

5.5.2 *Pandas*

Pandas is an open-source package designed for high-performance data manipulation and analysis in Python. Pandas provides powerful tools for "Data Cleaning and Transformation" processes dealing with missing values, filtering rows, and applying various operations such as reshaping. Furthermore, this library gives the chance not only to select subsets of data using labels (column names) or integer-based indexing, but also to easily group data based on specific criteria and compute summary statistics (mean, sum, etc.).

   Pandas is really powerful for its advanced data structures:

- DataFrame. It represents a two-dimensional, size-mutable, and heterogeneous tabular data structure with labeled axes (rows and columns) and it may contain different data type;
- Series. A one-dimensional labeled array—like a single column—that can hold any data type, but only one type inside. They are considered the building blocks of DataFrames.

Among other things, Pandas excels at handling time series of data, and dataframes can be joined and merged by considering columns or indexes in common.

Pandas can be easily installed using pip. Open your command-line interface or terminal and run the command

```
pip install pandas
```

Code 5.117 Pandas installation via `pip`

waiting for the installation to complete.

To begin with using Pandas, we propose a simple code to create a Series starting from data (i.e., "names" and "ages"):

```python
import pandas as pd

# Data
names = ['Sebastiano', 'Giuseppe', 'Ulderico']
ages = [36, 48, 64]

# Create a Series
name_series = pd.Series(ages, index=names)

# Display the Series
print("Name and Age Series:")
print(name_series)
# Output: Name and Age Series:
#         Sebastiano      36
#         Giuseppe        48
#         Ulderico        64
#         dtype: int64
```

Code 5.118 Simple series in Pandas

As already seen in the subsection devoted to NumPy, after installing Pandas (e.g., via "pip" as showed in the command line 5.117), it may be imported into Python code as reported in Line 1 of Code 5.120.[10] So, we use the `pd.Series()` function to create a Series where the names are used as the index and the ages as the data.

Starting from the data used in the previous example, we can construct a Dataframe by using the `pd.Dataframe()` function:

```python
import pandas as pd

# Create a simple DataFrame
data = {'names': ['Sebastiano', 'Giuseppe', 'Ulderico'],
        'ages': [36, 48, 64]}
df = pd.DataFrame(data)
print(df)
# Output:      names     ages
# 0       Sebastiano      36
```

---

[10] Note that (`import pandas as pd`) is only a common convention that allows the access to Pandas features with a short and recognizable prefix (`pd`).

```
10  # 1            Giuseppe      48
11  # 2            Ulderico      64
```

Code 5.119 Simple dataframe creation in Pandas

In this way, unlike the Series, we can index both rows and columns in the Dataframe named "df" to realize something like a table that can be really useful, for example, to plot data later. An important note is that, in addition to the presence of the columns with data that are represented by well-defined labels (i.e., "names" and "ages"), we find the first column with no label and it represents the index of the dataframe. In matrix/table style, it allows us to identify the rows of our dataframe.

Analysis tasks often begin with loading data from a file. Pandas seems to be one of the simplest libraries in doing this using the `.read_csv()` module allowing to read a file. In this frame, we may call the file '...\hooke.csv'—as shown in Example 5.120—with the goal of analyzing and displaying data obtained from experiments involving the elongation of springs hung vertically and subjected to different masses.

```python
 1  import pandas as pd
 2
 3  # Read a .csv file
 4  data = pd.read_csv('...\hooke.csv')
 5  print(data.info())
 6  # Output: <class 'pandas.core.frame.DataFrame'>
 7  #          RangeIndex: 10 entries, 0 to 9
 8  #          Data columns (total 4 columns):
 9  #           #   Column            Non-Null Count   Dtype
10  #          ---  ------            --------------   -----
11  #           0   Index             10 non-null      int64
12  #           1   "Mass (kg)"       10 non-null      float64
13  #           2   "Spring 1 (m)"    10 non-null      float64
14  #           3   "Spring 2 (m)"    10 non-null      float64
15  #          dtypes: float64(3), int64(1)
16  #          memory usage: 448.0 bytes
17  #          None
18
19  print(data)
20  # Output:   Index    "Mass (kg)"    "Spring 1 (m)"    "Spring 2 (m
        )"
21  #        0       1          0.00            0.050
        0.050
22  #        1       2          0.49            0.066
        0.066
23  #        2       3          0.98            0.087
        0.080
24  #        3       4          1.47            0.116
        0.108
25  #        4       5          1.96            0.142
        0.138
26  #        5       6          2.45            0.166
        0.158
27  #        6       7          2.94            0.193
        0.174
```

```
28 #        7          8          3.43              0.204
          0.192
29 #        8          9          3.92              0.226
          0.205
30 #        9         10          4.41              0.238
          0.232
```

Code 5.120 Read a datafile in Pandas with the .read_csv() module

Here, we propose a simple physics experiment regarding the measurements of the elongation of two different springs with several masses. In Example 5.120, we introduce the .info() method used for getting information about the number of rows, columns, memory usage, and count of not null.

Note that if we are trying to read a large file, it could be really useful to employ the .head() method just showing the first 5 lines of our data.

```python
1  import pandas as pd
2
3  # Read a .csv file
4  data = pd.read_csv('...\hooke.csv')
5  print(data.head())
6  # Output:   Index    "Mass (kg)"    "Spring 1 (m)"    "Spring 2 (m
       )"
7  #        0          1          0.00              0.050
          0.050
8  #        1          2          0.49              0.066
          0.066
9  #        2          3          0.98              0.087
          0.080
10 #        3          4          1.47              0.116
          0.108
11 #        4          5          1.96              0.142
          0.138
```

Code 5.121 Read a datafile in Pandas and make use of the .head() method

Another important feature of Pandas is to add new columns to existing data and then save the data to a new file. Several methods exist to achieve these goals; here we show only a few examples. Referring to our previous example on the stretching of springs, we might want to measure the elongation of another spring and then subsequently add the results of the measurement to the existing dataframe "data" by means of the .insert() method. After that, we use the .to_csv method to save the data in a new datafile (e.g., 'hooke_new.txt').

```python
1  import pandas as pd
2
3  # Read a .csv file
4  data = pd.read_csv('...\hooke.csv')
5
6  # Using DataFrame.insert() to add a column
7  data.insert(4, "Spring 3 (m)", [0.060, 0.074, 0.088, 0.116,
       0.146, 0.166, 0.182, 0.200, 0.213, 0.240], True)
```

```
 8
 9  #Save the data with .to_csv()
10  data.to_csv('hooke_new.txt')
```

Code 5.122 Read a datafile in Pandas, add a new column and save the data in a new datafile

Another method consists of the use of dictionaries. We can expand an existing dataframe (e.g., "data" in Example 5.120) with the data of columns "Spring 3 (m)" and "Spring 4 (m)" by using `.assign()`.

```
 1  import pandas as pd
 2
 3  # Read a .csv file
 4  data = pd.read_csv('...\hooke.csv')
 5
 6  # Define new data for additional columns
 7  spring_3 = [0.060, 0.074, 0.088, 0.116, 0.146, 0.166, 0.182,
         0.200, 0.213, 0.240]
 8  spring_4 = [0.061, 0.073, 0.084, 0.115, 0.141, 0.161, 0.187,
         0.205, 0.211, 0.246]
 9
10  # Add multiple columns using dictionary assignment
11  new_data = {'"Spring 3 (m)"': spring_3, '"Spring 4 (m)"':
         spring_4 }
12  data = data.assign(**new_data)
```

Code 5.123 Read a datafile in Pandas and add multiple columns by means of the `.assign()` method

5.5.2.1 Data Aggregation

As already seen in NumPy (see Sect. 5.5.1.1), data aggregation is a powerful way to summarize and analyze data. In Pandas, we may use many aggregation functions to object or numeric data in a DataFrame or Series, such as in Table 5.4.

In Pandas, we have several procedures—really useful even to descriptive statistics —that can be used to perform data aggregation, but here we show the `describe()` and `aggregate()` methods.

Referring to the first method, `.describe()` generates summary/basic statistics about the numerical and categorical different columns within a DataFrame. For numeric data, it provides mean, standard deviation, quartiles, minimum, maximum, and more. For object data (e.g., strings), it includes count, unique values, most common value (top), and its frequency (freq).[1] Let's see how to implement this function inside the code of the Example 5.120:

```
 1  import pandas as pd
 2
 3  # Read a .csv file
 4  data = pd.read_csv('...\hooke.csv')
 5
```

```
6  # Get summary statistics
7  print(data.describe())
8  # Output:     Index    "Mass (kg)"    "Spring 1 (m)"    "Spring 2 (
       m)"
9  # count  10.00000    10.000000       10.000000
       10.000000
10 # mean     5.50000     2.205000        0.148800
       0.140300
11 # std      3.02765     1.483549        0.067307
       0.062429
12 # min      1.00000     0.000000        0.050000
       0.050000
13 # 25%      3.25000     1.102500        0.094250
       0.087000
14 # 50%      5.50000     2.205000        0.154000
       0.148000
15 # 75%      7.75000     3.307500        0.201250
       0.187500
16 # max     10.00000     4.410000        0.238000
       0.232000
```

Code 5.124 Simple dataframe in Pandas and the `.describe()` method

Concerning the `aggregate()` method and taking Example 5.120 and its data again, we can write a code that allows us to use many of the aggregation functions just mentioned in Table 5.4 to perform a simple statistical analysis.

```
1  import pandas as pd
2
3  # Read a .csv file
4  data = pd.read_csv('...\hooke.csv')
5
```

Table 5.4 Description of some aggregation functions

Function	Description
count()	Total number of items
first()	First item
last()	Last item
mean()	Mean
median()	Median
min()	Minimum
max()	Maximum
var()	Variance
std()	Standard deviation
prod()	Product of all items
sum()	Sum of all items
cumsum()	Cumulative sum
cumprod()	Cumulative product

```
6 aggr_functions = data.aggregate({' "Mass (kg)"': ['count', '
      mean', 'median', 'min', 'max', 'var', 'std'], ' "Spring 1
      (m)"': ['count', 'mean', 'median', 'min', 'max', 'var', '
      std'], ' "Spring 2 (m)"': ['count', 'mean', 'median', 'min
      ', 'max', 'var', 'std']})
7 print(aggr_functions)
8 # Output:      "Mass (kg)"      "Spring 1 (m)"      "Spring 2 (m)"
9 # count       10.000000           10.000000           10.000000
10 # mean         2.205000            0.148800            0.140300
11 # median       2.205000            0.154000            0.148000
12 # min          0.000000            0.050000            0.050000
13 # max          4.410000            0.238000            0.232000
14 # var          2.200917            0.004530            0.003897
15 # std          1.483549            0.067307            0.062429
```

Code 5.125 Aggregation functions in Pandas

In this way, through a few lines of code and the combination of Pandas with aggregation functions, we are performing a basic statistical study of data acquired from a physics experiment.

5.5.2.2 Data Cleaning

The Pandas library offers a convenient method for eliminating unnecessary columns or rows from a DataFrame using the `drop()` function. To illustrate this, let's examine a straightforward example starting from the "data" used in Example 5.120). It is well evident that the presence of the "Index" column is not necessary for our purposes and can be removed.

```
1 import pandas as pd
2
3 # Read a .csv file
4 data = pd.read_csv('...\hooke.csv')
5
6 print("Original DataFrame:")
7 print(data)
8 # Output: Original DataFrame:
9 #    Index   "Mass (kg)"     "Spring 1 (m)"     "Spring 2 (m)"
10 # 0     1          0.00             0.050              0.050
11 # 1     2          0.49             0.066              0.066
12 # 2     3          0.98             0.087              0.080
13 # 3     4          1.47             0.116              0.108
14 # 4     5          1.96             0.142              0.138
15 # 5     6          2.45             0.166              0.158
16 # 6     7          2.94             0.193              0.174
17 # 7     8          3.43             0.204              0.192
18 # 8     9          3.92             0.226              0.205
19 # 9    10          4.41             0.238              0.232
20
21 # Drop a column
22 data_dropped_column = data.drop('Index', axis=1)
```

```
23 print("\nDataFrame after dropping 'Index' column:")
24 print(data_dropped_column)
25 # Output: DataFrame after dropping 'Index' column:
26 #        "Mass (kg)"      "Spring 1 (m)"      "Spring 2 (m)"
27 # 0         0.00               0.050               0.050
28 # 1         0.49               0.066               0.066
29 # 2         0.98               0.087               0.080
30 # 3         1.47               0.116               0.108
31 # 4         1.96               0.142               0.138
32 # 5         2.45               0.166               0.158
33 # 6         2.94               0.193               0.174
34 # 7         3.43               0.204               0.192
35 # 8         3.92               0.226               0.205
36 # 9         4.41               0.238               0.232
```

Code 5.126   Eliminate a label with the `.drop()` module

In the line of code #22 of Example 5.126, we employ the .drop() function inserting as parameters:

- label = 'Index', referring to the name of the column to drop;
- axis = 1, to drop labels from the index (0 or 'index') or columns (1 or 'columns').

In the example given, however, there is also another possibility to remove the 'Index' column and turn it into the "real" index of our dataframe. This can be done with the `.set_index()` module.

```
 1 import pandas as pd
 2
 3 # Read a .csv file
 4 data = pd.read_csv('...\hooke.csv')
 5
 6 print("Original DataFrame:")
 7 print(data)
 8 # Output: Original DataFrame:
 9 #       Index    "Mass (kg)"      "Spring 1 (m)"      "Spring 2 (m)"
10 # 0       1         0.00               0.050               0.050
11 # 1       2         0.49               0.066               0.066
12 # 2       3         0.98               0.087               0.080
13 # 3       4         1.47               0.116               0.108
14 # 4       5         1.96               0.142               0.138
15 # 5       6         2.45               0.166               0.158
16 # 6       7         2.94               0.193               0.174
17 # 7       8         3.43               0.204               0.192
18 # 8       9         3.92               0.226               0.205
19 # 9      10         4.41               0.238               0.232
20
21 # Drop the 'Index' column and set it as the index of our
       dataframe
22 data_set_index = data.set_index('Index')
23 print("\nDataFrame after dropping 'Index' column and set it as
       the index of our dataframe:")
24 print(data_set_index)
```

```
25 # Output: DataFrame after dropping 'Index' column and set it
      as the index of our dataframe:
26 #           "Mass (kg)"    "Spring 1 (m)"    "Spring 2 (m)"
27 # Index
28 # 1              0.00            0.050             0.050
29 # 2              0.49            0.066             0.066
30 # 3              0.98            0.087             0.080
31 # 4              1.47            0.116             0.108
32 # 5              1.96            0.142             0.138
33 # 6              2.45            0.166             0.158
34 # 7              2.94            0.193             0.174
35 # 8              3.43            0.204             0.192
36 # 9              3.92            0.226             0.205
37 # 10             4.41            0.238             0.232
```

Code 5.127 Set the index of a dataframe with the `.set_index()` module

5.5.2.3   Data Transformation

In this section, we will provide some useful methods to perform data transformation techniques (e.g., normalization) with Pandas.

One of the most used ways to convert data is to employ the `.transform()` method that calls function(s)—such as `np.sqrt()`, `np.exp()`, etc.—and/or performs mathematical operations, conversions, etc. on self, producing a DataFrame with the same axis shape as self. Let's say we have to carry out a physics experiment in which we want to measure the temperature of a substance at different time intervals, and we want to convert these temperatures from Celsius to Fahrenheit. Applying the `.transform()` method, we converted the temperatures from Celsius to Fahrenheit. This allows us to see how the temperature changes over time in both units, which can be useful for comparing data in different temperature scales. Let's see how to implement a code performing this kind of study in Pandas.

```python
import pandas as pd

# Create a sample DataFrame
data = {
    'Time (minutes)': [0, 10, 20, 30, 40],
    'Temperature (C)': [20, 25, 30, 35, 40]
}
df = pd.DataFrame(data)

# Define a function to convert Celsius to Fahrenheit
def celsius_to_fahrenheit(celsius):
    return (celsius * 9/5) + 32

# Use .transform() to apply the conversion function to the
    temperature data
df['Temperature (F)'] = df['Temperature (C)'].transform(
    celsius_to_fahrenheit)
```

```
16
17 print(df)
18 # Output:   Time (minutes)   Temperature (C)   Temperature (F)
19 # 0                      0                20                68.0
20 # 1                     10                25                77.0
21 # 2                     20                30                86.0
22 # 3                     30                35                95.0
23 # 4                     40                40               104.0
```

Code 5.128 The .transform() module in Pandas

We can also use Pandas to produce standardized, scaled, and normalized data as already seen in NumPy (see Sect. 5.5.1.3). In this frame, we realize the following example that contains the procedures to obtain the rescaled data.

```
1  import pandas as pd
2
3  # Read a .csv file
4  data = pd.read_csv('hooke.csv')
5
6  # Drop the 'Index' column and set it as the index of our
       dataframe
7  data = data.set_index('Index')
8
9  # Duplicate the original DataFrame
10 data_stand = data.copy()
11 data_scal = data.copy()
12 data_norm = data.copy()
13
14 # apply normalization techniques
15 # standardization
16 for column in data_stand.columns:
17     data_stand[column] = (data_stand[column] -
18                           data_stand[column].mean()) /
       data_stand[column].std()
19 # scaling in range [-1, 1]
20 for column in data_scal.columns:
21     data_scal[column] = data_scal[column]/ data_scal[column].
       abs().max()
22 # normalization
23 for column in data_norm.columns:
24     data_norm[column] = (data_norm[column] - data_norm[column
       ].min()) / (data_norm[column].max() - data_norm[column].
       min())
25
26 # Data comparison
27 print("Original DataFrame:")
28 print(data)
29 # Output:  Original DataFrame:
30 #           "Mass (kg)"    "Spring 1 (m)"    "Spring 2 (m)"
31 # Index
32 # 1               0.00            0.050             0.050
33 # 2               0.49            0.066             0.066
34 # 3               0.98            0.087             0.080
```

```
35 # 4                    1.47              0.116               0.108
36 # 5                    1.96              0.142               0.138
37 # 6                    2.45              0.166               0.158
38 # 7                    2.94              0.193               0.174
39 # 8                    3.43              0.204               0.192
40 # 9                    3.92              0.226               0.205
41 # 10                   4.41              0.238               0.232

43 print("\nDataFrame with standardized data:")
44 print(data_stand)
45 # Output: DataFrame with standardized data:
46 #          "Mass (kg)"     "Spring 1 (m)"     "Spring 2 (m)"
47 # Index
48 # 1        -1.486301       -1.467910          -1.446450
49 # 2        -1.156012       -1.230191          -1.190157
50 # 3        -0.825723       -0.918186          -0.965902
51 # 4        -0.495434       -0.487322          -0.517390
52 # 5        -0.165145       -0.101030          -0.036842
53 # 6         0.165145        0.255547           0.283523
54 # 7         0.495434        0.656696           0.539816
55 # 8         0.825723        0.820128           0.828145
56 # 9         1.156012        1.146990           1.036382
57 # 10        1.486301        1.325279           1.468875

59 print("\nDataFrame with scaled data:")
60 print(data_scal)
61 # Output: DataFrame with scaled data:
62 #          "Mass (kg)"     "Spring 1 (m)"     "Spring 2 (m)"
63 # Index
64 # 1         0.000000        0.210084           0.215517
65 # 2         0.111111        0.277311           0.284483
66 # 3         0.222222        0.365546           0.344828
67 # 4         0.333333        0.487395           0.465517
68 # 5         0.444444        0.596639           0.594828
69 # 6         0.555556        0.697479           0.681034
70 # 7         0.666667        0.810924           0.750000
71 # 8         0.777778        0.857143           0.827586
72 # 9         0.888889        0.949580           0.883621
73 # 10        1.000000        1.000000           1.000000

75 print("\nDataFrame with normalized data:")
76 print(data_norm)
77 # Output: DataFrame with normalized data:
78 #          "Mass (kg)"     "Spring 1 (m)"     "Spring 2 (m)"
79 # Index
80 # 1         0.000000        0.000000           0.000000
81 # 2         0.111111        0.085106           0.087912
82 # 3         0.222222        0.196809           0.164835
83 # 4         0.333333        0.351064           0.318681
84 # 5         0.444444        0.489362           0.483516
85 # 6         0.555556        0.617021           0.593407
86 # 7         0.666667        0.760638           0.681319
87 # 8         0.777778        0.819149           0.780220
```

```
88  # 9                0.888889              0.936170              0.851648
89  # 10               1.000000              1.000000              1.000000
```

Code 5.129 Standardization, scaling, and normalization processes in Pandas

Note that we make use of the ` .copy() ` function to duplicate the original Dataframe and, after performing the calculations on this replica, we print both the original and transformed data to show the differences between them.

The Pandas library includes numerous built-in methods for calculating the most common descriptive statistical functions, making data normalization techniques easy to implement.

5.5.2.4 Data Filtering

Data filtering in Pandas is essential for data analysis and manipulation. It allows you to focus on specific subsets of your data, making it easier to perform detailed analyses and derive insights. Whether you are filtering based on column values, multiple conditions, or using advanced methods like regular expressions, Pandas provides a versatile set of tools to handle your data efficiently.

Starting from Example 5.127, we can easily filter the data to consider all measurements involving masses from 1 to 4 kg. To do that, we can filter rows combining multiple conditions using the & (and) operator.

```python
import pandas as pd

# Read a .csv file
data = pd.read_csv('...\hooke.csv')

# Drop the 'Index' column and set it as the index of our
    dataframe
data_set_index = data.set_index('Index')

# Filter the data considering all the mass values between 1
    and 4
row_filter = data_set_index[
    ((data_set_index[' "Mass (kg)"'] >= 1) & (data_set_index['
     "Mass (kg)"'] <= 4))
    ]
print(row_filter)
# Output: "Mass (kg)"     "Spring 1 (m)"     "Spring 2 (m)"
# Index
# 4                1.47              0.116              0.108
# 5                1.96              0.142              0.138
# 6                2.45              0.166              0.158
# 7                2.94              0.193              0.174
# 8                3.43              0.204              0.192
# 9                3.92              0.226              0.205
```

Code 5.130 Filtering rows in Pandas

It is possible to filter columns by specifying a list of column names, as shown in Example 5.131.

```python
import pandas as pd

# Read a .csv file
data = pd.read_csv('...\hooke.csv')

# Drop the 'Index' column and set it as the index of our
    dataframe
data_set_index = data.set_index('Index')

# Select only the ' "Spring 1 (m)"' and ' "Spring 2 (m)"'
    columns
columns_filter = data_set_index[[' "Spring 1 (m)"', ' "Spring
    2 (m)"']]
print(columns_filter)
# Output:   "Spring 1 (m)"    "Spring 2 (m)"
# Index
# 1                  0.050             0.050
# 2                  0.066             0.066
# 3                  0.087             0.080
# 4                  0.116             0.108
# 5                  0.142             0.138
# 6                  0.166             0.158
# 7                  0.193             0.174
# 8                  0.204             0.192
# 9                  0.226             0.205
# 10                 0.238             0.232
```

Code 5.131 Filtering columns in Pandas

Example 5.131 can be easily reproduced by means of the `.filter()` method that can be used to subset rows or columns based on labels.

```python
import pandas as pd

# Read a .csv file
data = pd.read_csv('...\hooke.csv')

# Drop the 'Index' column and set it as the index of our
    dataframe
data_set_index = data.set_index('Index')

# Select columns that contain the sentence 'Spring'
filtered_data = data_set_index.filter(like='Spring', axis=1)
print(filtered_data)
# Output:   "Spring 1 (m)"    "Spring 2 (m)"
# Index
# 1                  0.050             0.050
# 2                  0.066             0.066
# 3                  0.087             0.080
# 4                  0.116             0.108
# 5                  0.142             0.138
# 6                  0.166             0.158
```

```
20 #  7                      0.193              0.174
21 #  8                      0.204              0.192
22 #  9                      0.226              0.205
23 #  10                     0.238              0.232
```

Code 5.132 The `.filter()` method in Pandas

Furthermore, we can easily access each record in our DataFrame using the `.loc[]` class instance. Despite its somewhat unintuitive name, it enables us to reference a row or record by its label rather than its position (*label-based indexing*). For instance, referring to the Example 5.127, 1 is the first label in the index, so we may use .loc[1] if we want to call only the first row of data acquired of our experiment (i.e., the measurements with mass (kg) equal to 0). To access the same row by its position, we would use `.iloc[0]`, which performs *position-based indexing*.

```python
import pandas as pd

# Read a .csv file
data = pd.read_csv('...\hooke.csv')

# Drop the 'Index' column and set it as the index of our
    dataframe
data_set_index = data.set_index('Index')
print("\nDataFrame after dropping 'Index' column and set it as
    the index of our dataframe:")
print(data_set_index)

print("\nAccess a record:")
print(data_set_index.loc[1])
# Output: "Mass (kg)"           0.00
#         "Spring 1 (m)"        0.05
#         "Spring 2 (m)"        0.05
print(data_set_index.iloc[0])
# Output: "Mass (kg)"           0.00
#         "Spring 1 (m)"        0.05
#         "Spring 2 (m)"        0.05
```

Code 5.133 Access a record of a dataframe with the `.loc[]` and `.iloc[]` modules

Note that both command lines #25 and #29 of Example 5.133 print the data contained in the first line of our dataframe.

5.5.2.5 Summary

Pandas is an essential tool for data scientists, analysts, and anyone working with structured data. Its flexibility and ease of use make it a go-to library for data manipulation and exploration. Pandas is well supported and well documented, so you will be able to find plenty of examples to help you out. Here are a few guides to help get you started:

- http://pandas.pydata.org/pandas-docs/dev/basics.html

- http://pandas.pydata.org/pandas-docs/dev/10min.html
- http://pandas.pydata.org/pandas-docs/dev/cookbook.html#cookbook
- http://www.gregreda.com/2013/10/26/working-with-pandas-dataframes/.

5.5.3 *Matplotlib*

Matplotlib is a comprehensive library for creating static, animated, and interactive visualizations in Python. It provides a quick way to visualize data from Python and a wide range of plotting functions, from histograms, scatter plots, and bar charts to error charts, power spectra, and heatmaps. It allows the customization of every aspect of a figure, including size, DPI, line width, color, style, axes, and more. These features make it suitable for various scientific and publication-quality graphics.

As usual, we have several ways to install Matplotlib, here we report the installation via `pip`:

```
python -m pip install -U matplotlib
```
Code 5.134 Matplotlib installation via `pip`

"matplotlib.pyplot" is a collection of command-style functions that permit making changes to a figure (e.g., creates a figure, creates a plotting area in a figure, plots some lines in a plotting area, decorates the plot with labels, etc.), providing a procedural interface to the matplotlib object-oriented plotting library. We can call this submodule by typing in our code the following statements:

```
import matplotlib.pyplot as plt
```
Code 5.135 Import pyplot in Python

or

```
from matplotlib import pyplot as plt
```
Code 5.136 Import pyplot in Python

Note that pyplot is usually imported under the `plt` alias.

Matplotlib is designed to work with NumPy arrays and is often used alongside libraries like Pandas for broader data analysis tasks. The combined use of NumPy and Matplotlib allows for creating graphs and visual representations of data arrays. In the following Example 5.137, we want to draw the sine function, making use of the command `.plot()` and starting from the default settings. The first step is to create a Python script in which we write our code. Import NumPy and matplotlib.pyplot as indicated in 5.137 and then create our data ("X") with the command `np.linspace(-np.pi, np.pi, 128)`, which returns 128 evenly spaced samples, calculated over the interval $-\pi$ (`-np.pi`) and π (`np.pi`). The sine function is calculated by means of `np.sin(X)` and then stored in the variable "S". Lines

Fig. 5.60 Visualization with Matplotlib. The output of the script is illustrated in the Example 5.137

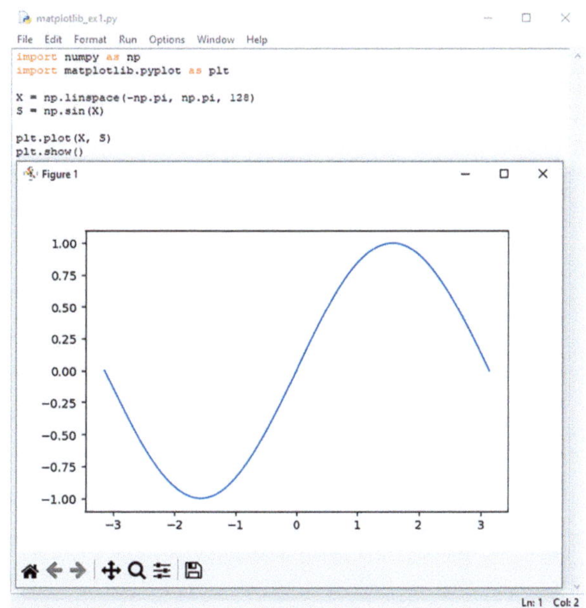

plt.plot(X, S) and plt.show() are employed to generate a line chart, plotting "X" along the x-axis and "S" along the y-axis, and to show it, respectively.

```python
import numpy as np
import matplotlib.pyplot as plt

X = np.linspace(-np.pi, np.pi, 128)
S = np.sin(X)

# Plot sine
plt.plot(X, S)
# Show result on screen
plt.show()
```

Code 5.137   Example of combining NumPy and Matplotlib in Python

Figure 5.60 shows the script illustrated in the Example 5.137 and executed in Python's IDLE, reporting the graph produced by using matplotlib.plt.

Of course, such code can be improved by customizing the plot to our liking (e.g., decorating the plot with labels). Referring to the Example 5.137, let's see how instantiating all the figure settings that influence the appearance of the plot. First of all, consider the following code:

```python
import numpy as np
import matplotlib.pyplot as plt

# Create a figure of size 8x6 inches, 300 dots per inch
plt.figure(figsize=(8, 6), dpi=300)

```

```python
7  # Create a new subplot from a grid of 1x1
8  plt.subplot(1, 1, 1)
9
10 X = np.linspace(-np.pi, np.pi, 128)
11 S = np.sin(X)
12
13 # Set fonts
14 font1 = {'family':'serif','color':'blue','size':20}
15 font2 = {'family':'serif','color':'black','size':15}
16
17 # Set title
18 plt.title("Customize a plot", fontdict = font1, loc = 'right')
19
20 # Set x limits
21 plt.xlim(-4.0, 4.0)
22
23 # Set x ticks
24 plt.xticks(np.linspace(-4, 4, 9))
25
26 # Set x label
27 plt.xlabel("X", fontdict = font2)
28
29 # Set y limits
30 plt.ylim(-1.0, 1.0)
31
32 # Set y ticks
33 plt.yticks(np.linspace(-1, 1, 5))
34
35 # Set y label
36 plt.ylabel("sin(X)", fontdict = font2)
37
38 # Plot sine with a red continuous line of width 1 (pixels)
39 plt.plot(X, S, color="red", linewidth=1.0, linestyle="-",
             label="sine", marker = 'o')
40
41 # Text annotation inside the plot
42 plt.annotate('min', xy=(-1.6, -1), xytext=(1, -0.5),
43              arrowprops=dict(facecolor='black', shrink=0.05),
44              )
45
46 # Show legend
47 plt.legend(loc='upper left')
48
49 # Save figure using 300 dots per inch
50 plt.savefig("matplotlib_ex2.png", dpi=300)
51
52 # Show result on screen
53 plt.show()
```

Code 5.138 Customization of the figure setting with Matplotlib in Python

Fig. 5.61 Customization of the figure setting with Matplotlib in Python, as reported in Code 5.138

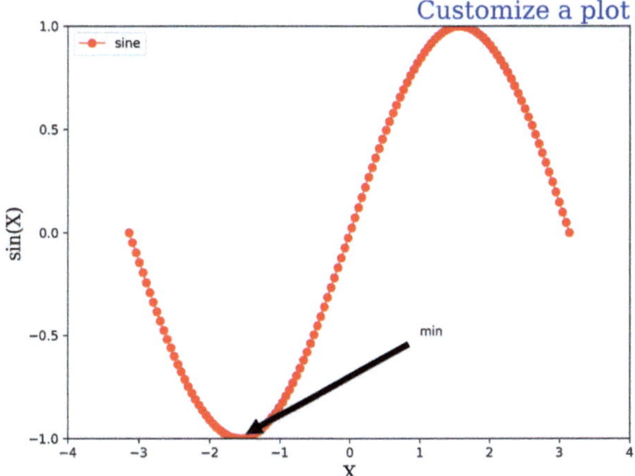

that produces this graph as a result:

It is well evident the difference between Figs. 5.60 and 5.61. Let's the a brief explanation of the Example 5.138.

So far—see Example 5.137—we employed figure and axes creation implicitly. This is useful when we want to make plots quickly, but it is not the preferred situation when we need to customize graphs or display them with much more detail. This is where additional functions such as   figure() ,   subplots() and   axes() come to

our aid.   plt.figure() in matplotlib allows the user to consider the whole window in the user interface, giving the possibility of dividing the plot into subunits called "subplots". In this frame, a figure exists as a window in the graphical user interface (GUI) with "Figure #" as title.   plt.subplots() is a function that returns a tuple

containing a figure and axes object(s). By using   plt.subplot() , we may arrange plots in a regular grid, specifying the number of rows and columns and the number of the plot.

The   axes() class is very useful because it allows placement of plots at any location inside the figure. It contains the plotted data, axis ticks, labels, title, legend, etc. Its methods are the main interface for manipulating the plot.

```python
import matplotlib.pyplot as plt
import numpy as np

# Some data to display
x = np.linspace(0, 3 * np.pi, 500)
y = np.sin(x ** 2)

fig, ax = plt.subplots()
ax.plot(x, y, color='green', marker='o', linestyle='dashed',
        linewidth=1, markersize=6)
ax.set_title('Single subplot()')
plt.show()
```

Code 5.139 Single subplots() in Python

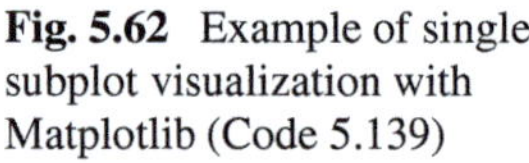

Fig. 5.62 Example of single subplot visualization with Matplotlib (Code 5.139)

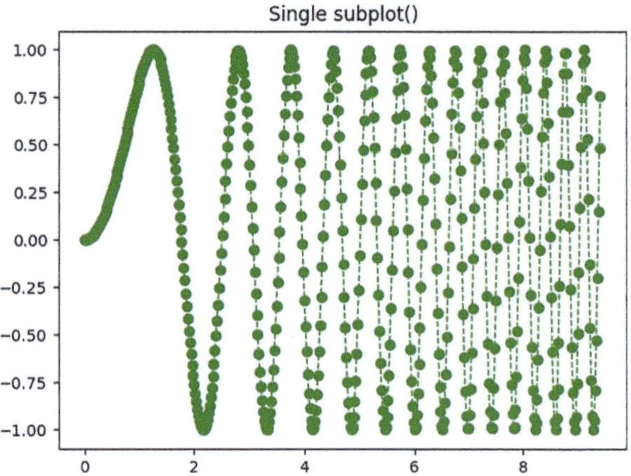

In this example, we highlight the fact that the use of `.subplots()` without arguments returns a `figure` and a single `axes`. As usual, after importing the libraries we need, we create some data and then calculate the sin function—in this case, we make use of NumPy's routines, i.e., `np.linspace()` and `np.sin()` — plotting the results in a graph by means of `.subplots()`. As already mentioned, `plt.subplots()` furnishes a tuple with a figure and axes object(s). In this frame, by using `fig, ax = plt.subplots()`, the outcome is the unpacking of this tuple into the variables `fig`[11] and `ax`, abbreviated in this way to represent `figure` and `axes`, respectively. By means of `ax.plot()` and `ax.set_title()`, we can graph y against x, choosing the title and the settings that we like best to customize the graph (e.g., color = 'green', marker = 'o', linestyle = 'dashed', linewidth = 1, markersize = 6). Finally, by running the code in IDLE, the command `plt.show()` gives Fig. 5.62 as result.

Subplot() allows you to make multiple graphs. So, referring to Example 5.139, we produce a vertically stacked subplot as shown in Example 5.140.

```python
import matplotlib.pyplot as plt
import numpy as np

# Some data to display
x = np.linspace(0, 3 * np.pi, 500)
```

---

[11] Note that considering `fig` as variable may be fruitful if we would like to change figure-level attributes or save the figure later (e.g. with `fig.savefig('yourfilename.png')`).

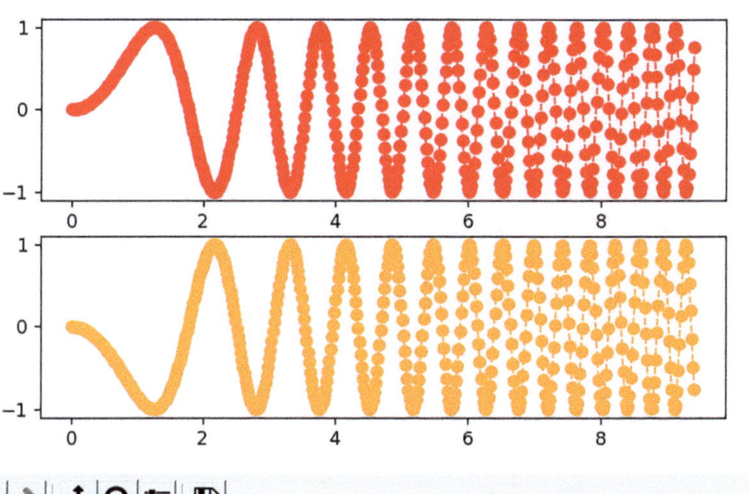


```python
import matplotlib.pyplot as plt
import numpy as np

# Some data to display
x = np.linspace(0, 3 * np.pi, 500)
y = np.sin(x ** 2)

fig, (ax1, ax2) = plt.subplots(2)
fig.suptitle('Multiple subplots')
ax1.plot(x, y, color='red', marker='o', linestyle='dashed', linewidth=1, markers
ax2.plot(x, -y, color='orange', marker='o', linestyle='dashed', linewidth=1, mar
plt.show()
```

Fig. 5.63 Example of multiple subplots as a results of executing Code 5.140

```python
 6  y = np.sin(x ** 2)
 7
 8  fig, (ax1, ax2) = plt.subplots(2)
 9  fig.suptitle('Multiple subplots')
10  ax1.plot(x, y, color='red', marker='o', linestyle='dashed',
        linewidth=1, markersize=6)
11  ax2.plot(x, -y, color='orange', marker='o', linestyle='dashed'
        , linewidth=1, markersize=6)
12  plt.show()
```

Code 5.140 Multiple subplots() in Python

Going back to Example 5.138, we made use of several functions in pyplot module of matplotlib library (Fig. 5.63):

- `plt.xlim()` and `plt.ylim()` get or set the limits of the current Axes;

- plt.xticks() and plt.yticks() get or set the current tick locations and labels of the x- and y-axis, respectively;
- plt.xlabel() and plt.ylabel() set the label for the x- and y-axis, respectively;
- in its simplest form, plt.annotate() allows the user to insert text on the point xy, but it also gives the possibility to display the text in another position with an arrow pointing from the text to the annotated point xy (as shown in the example);
- plt.legend() puts a legend on the Axes.
- plt.savefig() stores the current figure in a file (e.g., .svg, .png).

Matplotlib can also be used in conjunction with Pandas to plot data. Referring to what we have seen in Sect. 5.5.2, we can call up the Example 5.120 and implement the matplotlib code to complete it with data visualization:

```python
import pandas as pd
import matplotlib.pyplot as plt
import numpy as np

# Read a .csv file
data = pd.read_csv('...\hooke.csv')
data
# Output:   Index   "Mass (kg)"   "Spring 1 (m)"   "Spring 2 (m
    )"
#       0      1      0.00           0.050
    0.050
#       1      2      0.49           0.066
    0.066
#       2      3      0.98           0.087
    0.080
#       3      4      1.47           0.116
    0.108
#       4      5      1.96           0.142
    0.138
#       5      6      2.45           0.166
    0.158
#       6      7      2.94           0.193
    0.174
#       7      8      3.43           0.204
    0.192
#       8      9      3.92           0.226
    0.205
#       9     10      4.41           0.238
    0.232

# Plot data in the same graph
ax1 = plt.scatter(x = data[' "Mass (kg)"'], y=data[' "Spring 1
    (m)"'], label='Spring 1')
ax2 = plt.scatter(x = data[' "Mass (kg)"'], y=data[' "Spring 2
    (m)"'], label='Spring 2')

```

Fig. 5.64 Plot obtained from the combination of Matplotlib and Pandas as shown in Code 5.141

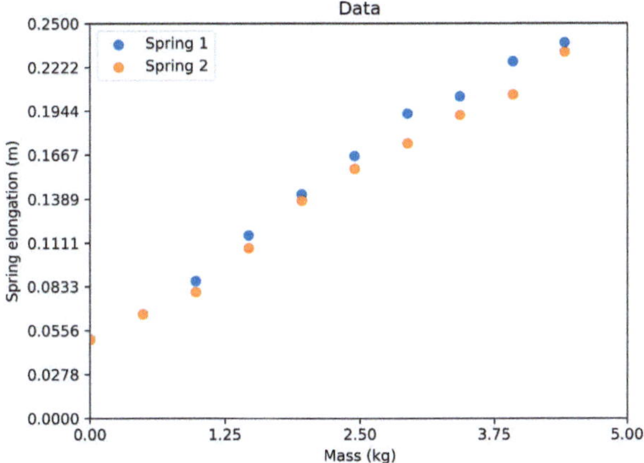


```python
24  # Set title
25  plt.title("Data")
26
27  # Set x limits
28  plt.xlim(0, 5)
29
30  # Set x ticks
31  plt.xticks(np.linspace(0, 5, 5))
32
33  # Set x label
34  plt.xlabel("Mass (kg)")
35
36  # Set y ticks
37  plt.yticks(np.linspace(0, 0.25, 10))
38
39  # Set y label
40  plt.ylabel("Spring elongation (m)")
41
42  # Show legend
43  plt.legend(loc='upper left')
44
45  # Save figure using 300 dots per inch
46  plt.savefig("pandas_matplotlib.png", dpi=300)
47
48  plt.show()
```

Code 5.141 Pandas and Matplotlib

After reading the file containing the data, we implement the code to display the data in the same final graph, saved by   plt.savefig()   and shown here in Fig. 5.64.

Note that   plt.scatter()   in Example 5.141 allows us to generate a customizable scatterplot of y vs x.

Matplotlib also allows data to be graphed in other ways, taking advantage of bar or pie plots. Let's see some examples of this type of chart. Imagine that you are a teacher having four classes (A, B, C, D) with a number of students in each class. We may employ the plt.bar() function to create the bar chart representing our data, customizing it with labels and a title.

```python
import matplotlib.pyplot as plt

# Data
classes = ['Class A', 'Class B', 'Class C', 'Class D']
students = [23, 17, 35, 29]

# Create bar chart
plt.figure(figsize=(10, 6))
plt.bar(classes, students, color='skyblue')
plt.xlabel('Classes')
plt.ylabel('Number of Students')
plt.title('Number of Students in Different Classes')
plt.show()
```

Code 5.142 Bar chart in Python

Referring to the same data used in Example 5.142, we may apply the plt.pie() function to create the pie chart, and then we customize it with labels, percentage display, and a title.

```python
import matplotlib.pyplot as plt

# Data
classes = ['Class A', 'Class B', 'Class C', 'Class D']
students = [23, 17, 35, 29]

# Create pie chart
plt.figure(figsize=(8, 8))
plt.pie(students, labels=classes, autopct='
plt.title('Student Distribution')
plt.show()
```

Code 5.143 Pie chart in Python

The "startangle" parameter rotates the start of the pie chart for better visualization.

5.5.3.1 Histograms

In physics, histograms are really important because of several key points:

1. they provide a clear visual representation of data distributions, making it easier to identify patterns, trends, and outliers;
2. they help in understanding the statistical properties of data, such as mean, variance, and standard deviation, which are crucial for experimental physics;
3. histograms allow physicists to compare different data sets effectively, facilitating the analysis of experimental results;
4. they can highlight anomalies or errors in data collection, ensuring the accuracy and reliability of experimental findings.

Matplotlib allows us to compute and plot a histogram by using the `plt.hist()` function. This method involves the use of several parameters, among which it is important to mention input values and bins (please refer to the web-reference of the plt.hist() function for more details).

Example #5.144 shows data of an experiment with the scope of measuring the velocities of a certain object (e.g., a material point in classical mechanics). After generating sample random data, we create the histogram for visualizing the distribution of velocities, which can help in understanding the spread and central tendency of the data.

```python
import matplotlib.pyplot as plt
import numpy as np

# Generate sample data: velocities of object (in m/s)
np.random.seed(0)
velocities = np.random.normal(loc=50, scale=10, size=1000)  #
    mean=50, std=10, n=1000

# Create histogram
plt.figure(figsize=(10, 6))
plt.hist(velocities, bins=30, color='blue', edgecolor='black')
plt.xlabel('Velocity (m/s)')
plt.ylabel('Frequency')
plt.title('Distribution of Velocities')
plt.grid(True)
plt.savefig("matplotlib_histogram.png", dpi=300)
plt.show()
```

Code 5.144 Histograms in Matplotlib

This example produces the histogram illustrated in Fig. 5.65.

Let us devote time now to a complete description of the code, particularly focusing our attention on functions and methods not described so far. Generating the sample data consists of using the following pieces of code:

- `np.random.seed(0)` (line #5) sets a number to start with (a seed value) for the random number generator to ensure reproducibility of the results.
- `velocities = np.random.normal(loc=50, scale=10, size=1000)` (line #6) generates an array of 1000 random numbers drawn from a normal (Gaussian) distribution with a mean ("loc") of 50 and a standard deviation ("scale") of 10. These numbers represent the velocities of the system under study in meters per second (m/s).

By using the `plt.hist(velocities, bins=30, color='blue', edgecolor='black')` command (line #10), we create a histogram of the "velocities" data with 30 bins. As shown in Fig. 5.65, the bars of the histogram are colored blue, and the edges of the bars are colored black. Note that a grid has been added to the graph for greater readability by line of code `plt.grid(True)`.

It may also happen that we find ourselves in a case where we have two velocity distributions referring to the same object, and we would like to see the two graphs

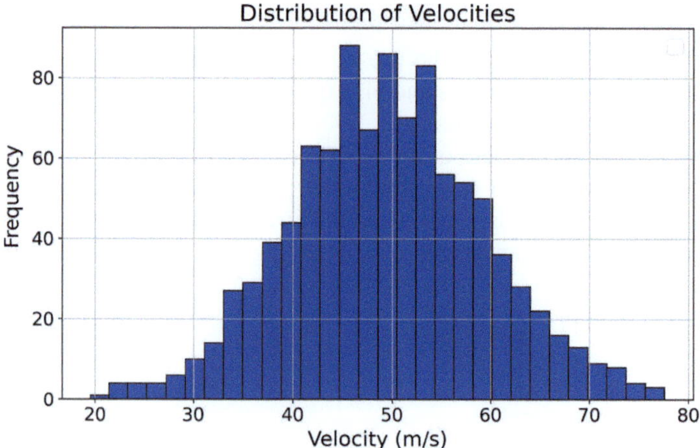

Fig. 5.65 Histogram in Python with Matplotlib representing the distribution of velocities of an object in m/s (Code 5.144)

simultaneously to make a quick comparison. It is therefore possible to create two histograms and place them side by side in the same graph using the plt.subplots() function as reported in the following example:

```python
import matplotlib.pyplot as plt
import numpy as np

# Generate sample data for two different sets of velocities
np.random.seed(0)
velocities1 = np.random.normal(loc=50, scale=10, size=1000)  #
    mean=50, std=10, n=1000
velocities2 = np.random.normal(loc=60, scale=15, size=1000)  #
    mean=60, std=15, n=1000

# Create subplots
fig, axs = plt.subplots(2, 1, figsize=(10, 12), sharex=True)

# First histogram
axs[0].hist(velocities1, bins=30, color='blue', edgecolor='
    black', label='Set 1')
axs[0].set_ylabel('Frequency')
axs[0].legend()
axs[0].grid(True)

# Second histogram
axs[1].hist(velocities2, bins=30, color='green', edgecolor='
    black', label='Set 2')
axs[1].set_xlabel('Velocity (m/s)')
axs[1].set_ylabel('Frequency')
axs[1].legend()
axs[1].grid(True)
```

```
24
25 # Add a single title for the entire figure
26 fig.suptitle('Comparison of Velocities', fontsize=16)
27
28 # Adjust layout
29 plt.tight_layout(rect=[0, 0, 1, 0.96])
30 plt.savefig("matplotlib_histograms-comp.png", dpi=300)
31 plt.show()
```

Code 5.145 Histograms and subplots in Matplotlib

that generates Fig. 5.66 as output. Let's look at the code used in Example 5.145. First, we import the libraries and then generate the sample data. Then, we create the figure with two subplots arranged vertically and with the same x-axis (i.e., "sharex = True"). This figure contains two histograms:

1. the first one that refers to the first set of velocities is generated by the line #13 in Example 5.145. This histogram has a blue color with black edges, and we set and add its label to distinguish it from the other histograms.
2. The second histogram plots the data of the second set of velocities (see line #19) with green color, black edges, and legend label 'Set 2'.

Note that both plots are characterized by having a bin equal to 30 in addition to the same x-scale for better comparison of the two data sets. Furthermore, we add the line plt.tight_layout(rect = [0, 0, 1, 0.96]) to adjust the layout, preventing overlap and ensuring the title is not cut off.

5.5.3.2 Errors Bar

An important role in experimental physics is given by errors because, in any experiment, we have to estimate the uncertainty of the measured values, which gives a range within which the true value of a physical quantity is expected to lie. Matplotlib gives the possibility to see the difference between the actual value and the calculated or measured value of a physical quantity via a graphical representation named "error bars". So, they give a general idea of how precise a measurement is, or conversely, how far from the reported value the true (error-free) value might be. We can add error bars to plots using the .errorbar() function. Below, Example 5.146 shows the measurements obtained in an experiment in which we were measuring force and acceleration—remember that the force exerted on a material point is directly proportional to its acceleration, according to Newton's second law of motion ($F = ma$).

```
1 import matplotlib.pyplot as plt
2 import numpy as np
3
4 # Data
5 acceleration = np.array([1, 2, 3, 4, 5])  # acceleration
      values
```

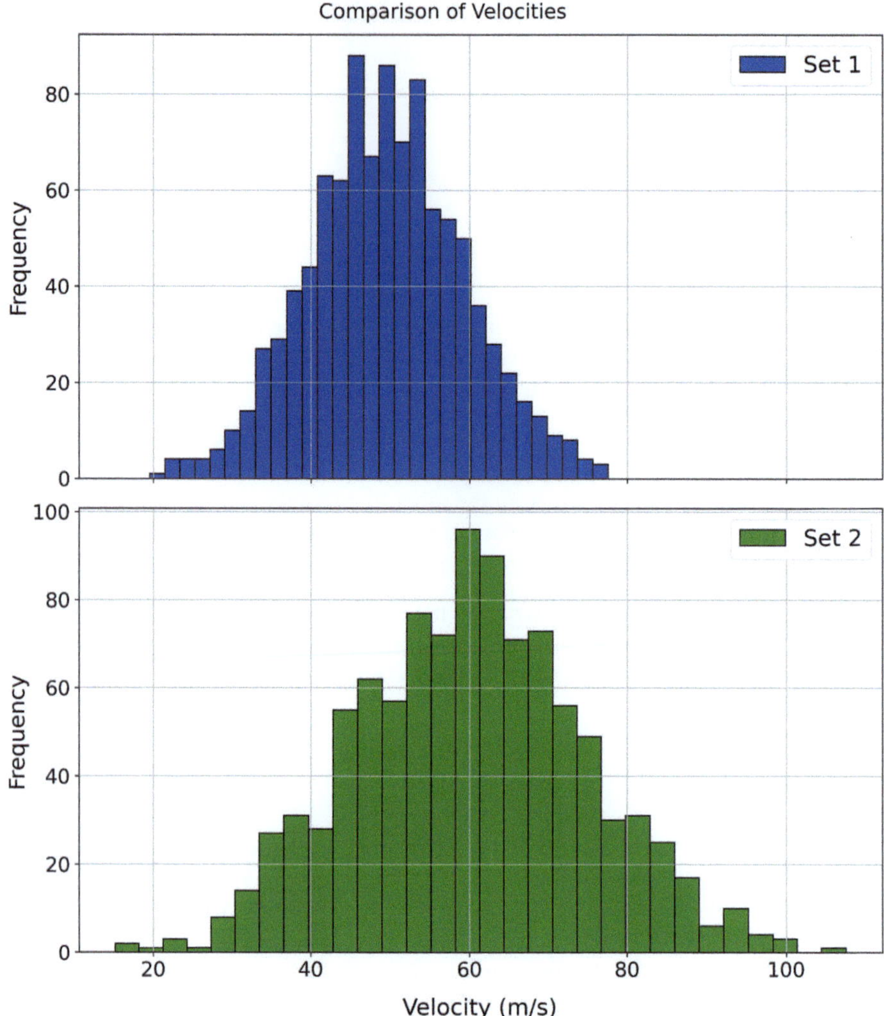

Fig. 5.66 Histograms generated by Code 5.145)

```python
force = np.array([11, 19, 31, 40, 52])  # force measurements

# Example variable error bar values
force_err = np.array([1, 2, 1.5, 3, 2.5])  # uncertainties in
    force measurements
acceleration_err = np.array([0.1, 0.2, 0.15, 0.3, 0.25])  #
    uncertainties in acceleration measurements

plt.figure()
plt.errorbar(acceleration, force, xerr=acceleration_err, yerr=
    force_err, fmt='o')
```

Fig. 5.67 Plot of force measurements with error bars generated by Code 5.146

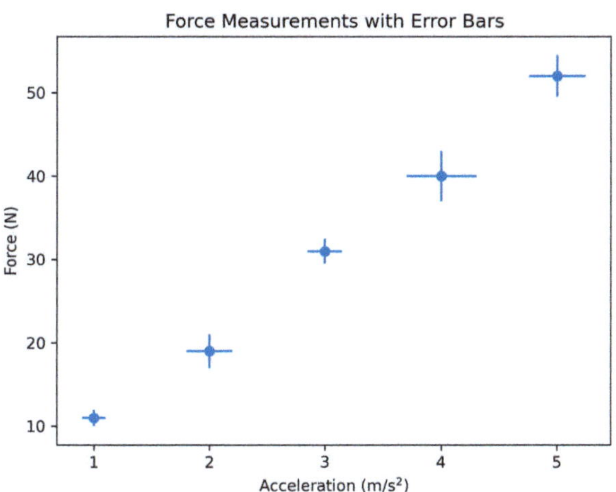


```
14  plt.xlabel(r'Acceleration (m/s²)')
15  plt.ylabel('Force (N)')
16  plt.title('Force Measurements with Error Bars')
17  plt.savefig("matplotlib_errors-bar.png", dpi=300)
18  plt.show()
```

Code 5.146 Errors bar in Matplotlib.

In this code, "acceleration" and "force" are our data points, "acceleration_err" ("xerr") and "force_err" ("yerr") are the error margins for the force and acceleration values, and fmt = 'o' specifies that the data points should be represented as dots. Note that it is possible to customize the appearance of the error bars, e.g., by using the "ecolor" parameter (changing the color of the error bars), or the "elinewidth" parameter (adjusting the width of the error bars).

Example 5.146 provides Fig. 5.67 as result.

5.5.3.3 Summary

Matplotlib is an essential tool for data visualization in Python, widely used in the scientific computing community. It is a very powerful library that contains many functions and allows extensive customization of a graph in its entirety. For more information, you can explore the official Matplotlib documentation.

5.5.4 SciPy

SciPy is an open-source Python library used for scientific and technical computing. It provides many user-friendly and efficient numerical routines, including statistics, optimization, and fitting modules, linear algebra, integration, interpolation, special functions, FFT, signal and image processing, ODE solvers, and more. Furthermore, SciPy builds on NumPy arrays as the basic data structure and is part of the NumPy stack which includes tools like Matplotlib and Pandas.

SciPy can be installed using several methods. With "pip" or Anaconda's conda, there is the chance to choose the best package version for a specific project, instead of system package managers, like "apt-get", installing SciPy across the entire computer, often have older versions or simply don't have as many available versions. As reported in the SciPy website, if you don't know which installation method you need or prefer, we recommend the Scientific Python Distribution Anaconda. Here, we report the installation of SciPy from PyPI with `pip`[12]:

```
python -m pip install scipy
```
Code 5.147 SciPy installation via `pip`

As already mentioned, SciPy is very useful for optimization problems that involve finding a numerical solution to a minimization or equality.

The scipy.optimize module provides algorithms for function minimization (scalar or multi-dimensional), curve fitting, and root finding. In physics, we usually make use of the graphical method with the purpose of visually representing and analyzing relationships between different variables in a given problem or experiment. As a matter of fact, it is possible to easily identify trends, patterns, and relationships between the variables by plotting data points on a graph.

In this frame, if we consider the Example 5.120 that generates the graph displayed in Fig. 5.64, we can try to use the curve fitting procedure to derive the value of the spring constant from Hooke's law. Starting from Hooke's law for a spring,

$$F_{hooke} = -kx, \tag{5.2}$$

where k is the spring constant and x is its extension, and writing the equilibrium relationship between the elastic force F_{hooke} and the weight force $P = mg$,

$$\begin{aligned} P &= F_{hooke} \Rightarrow \\ \Rightarrow mg &= kx \Rightarrow \\ \Rightarrow x &= \frac{g}{k}m, \end{aligned} \tag{5.3}$$

We can derive a relationship useful for tying the elongation of the spring with the mass hanging on the spring. The resulting equation is really simple and can be used to find the spring constants performing the "linear regression" method by means of the `scipy.stats.linregress()` function. Let's see how to implement this procedure with SciPy in our previous example.

```
import pandas as pd
```

---

[12] Note that installing SciPy automatically will also install NumPy..

```python
import matplotlib.pyplot as plt
import numpy as np
from scipy import stats

# Read a .csv file
data = pd.read_csv('hooke.csv')

x = data[' "Mass (kg)"']
y1 = data[' "Spring 1 (m)"']
y2 = data[' "Spring 2 (m)"']

result1 = stats.linregress(x, y1)
result2 = stats.linregress(x, y2)

slope1 = result1.slope
intercept1 = result1.intercept
fit1 = intercept1 + slope1 * x
slope2 = result2.slope
intercept2 = result2.intercept
fit2 = intercept2 + slope2 * x

print(f"R-squared for Spring #1: {result1.rvalue**2:.6f}")
print(f"R-squared for Spring #2: {result2.rvalue**2:.6f}")

# Output:
# R-squared for Spring #1: 0.991265
# R-squared for Spring #2: 0.992243

# Plot data in the same graph
ax1 = plt.scatter(x, y1, label='Spring 1')
ax2 = plt.plot(x, fit1, 'b', label='Fit - Spring 1')
ax3 = plt.scatter(x, y2, label='Spring 2')
ax4 = plt.plot(x, fit2, 'r', label='Fit - Spring 2')

k1 = 9.81/result1.slope
k2 = 9.81/result2.slope

print(f"Elastic Constant for Spring #1 (N/m): {k1:.6f}")
print(f"Elastic Constant for Spring #2 (N/m): {k2:.6f}")

# Output:
# Elastic Constant for Spring #1 (N/m): 217.179217
# Elastic Constant for Spring #2 (N/m): 234.033196

# Set title
plt.title("Data")

# Set x limits
plt.xlim(0, 5)

# Set x ticks
plt.xticks(np.linspace(0, 5, 5))
```

```python
55  # Set x label
56  plt.xlabel("Mass (kg)")
57
58  # Set y ticks
59  plt.yticks(np.linspace(0, 0.25, 10))
60
61  # Set y label
62  plt.ylabel("Spring elongation (m)")
63
64  # Show legend
65  plt.legend(loc='upper left')
66
67  # Text
68  plt.text(1.25, 0.03, f"Elastic Constant for Spring #1 (N/m): {
        k1:.6f}", color='b') # Coordinates and text
69  plt.text(1.25, 0.015, f"Elastic Constant for Spring #2 (N/m):
        {k2:.6f}", color='r') # Coordinates and text
70
71  # Save figure using 300 dots per inch
72  plt.savefig("lin_regr_scipy.png", dpi=300)
73
74  plt.show()
```

Code 5.148   linregress() function in SciPy

This code gives the graph illustrated in Fig. 5.68 as a result.

In Example 5.120, we have considered two different springs. Here, we want to calculate a linear least-squares regression for two sets of measurements and then compute the value of two spring constants. As a result, the   scipy.stats.linregress()   function returns an object with several information (floats), such as the   .slope ,

the   .intercept , and the   .rvalue   attributes:

- *slope* of the regression line;

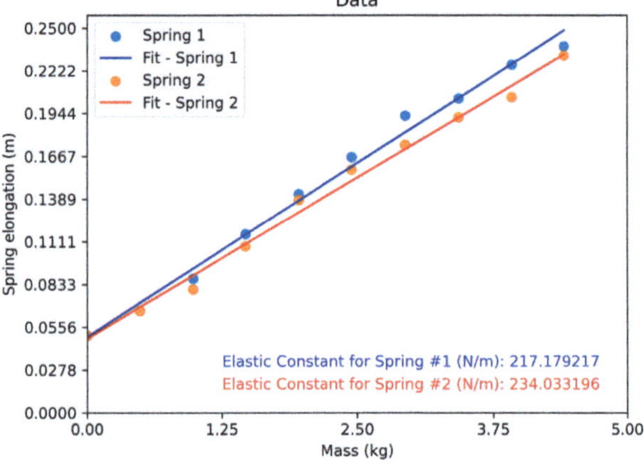

Fig. 5.68 Linear regression plot obtained with SciPy to calculate the spring constant from Hooke's law

- *intercept* of the regression line;
- *rvalue r* is the *Pearson correlation coefficient*, usually considered squared (r^2) as it represents the *coefficient of determination* used to evaluate the quality of fit of a model on data.[13]

These parameters allow us to obtain the regression lines (i.e., variables "fit1" and "fit2" reported in code #5.148), so it is possible to plot them as done in Fig. 5.68. Furthermore, by using the expression (5.3), we can roughly say that the value of the slope found for the two data sets is inversely proportional with the respective elastic constants. The computed values are thus reported as text in the graph generated using matplotlib—see Fig. 5.68.

Another example of employing SciPy regards the numerical integration of a function of a single variable by means of   scipy.integrate.quad() . This method computes a definite integral integrating a function from *a* to *b*.

```python
from scipy.integrate import quad

# Define a simple function for integration
def integrand(x):
    return x**3

# Perform numerical integration
result, error = quad(integrand, 0, 1)
print(f"Integral result: {result}, Error estimate: {error}")

# Output: Integral result: 0.25, Error estimate:
    2.7755575615628914e-15
```

Code 5.149   integrate.quad()   function in SciPy

Example 5.150 demonstrates using SciPy's "quad" function from the "integrate" module to perform numerical integration of a simple cubic function over the interval [0, 1].

Interpolation is one of the most important features of SciPy.   scipy.interpolate is really useful for fitting a function from experimental data and thus evaluating points where no measure exists. Here's a simple example of using   scipy.interpolate.interp1d()   for 1-D interpolation:

```python
import numpy as np
import matplotlib.pyplot as plt
from scipy import interpolate

# Given data points
x = np.arange(0, 10)
y = np.exp(-x / 3.0)

# Create an interpolation function
f = interpolate.interp1d(x, y)
```

---

[13] The coefficient of determination is a number between 0 and 1 that expresses what fraction of the variability of your dependent variable (Y) is explained by your independent variable (X).

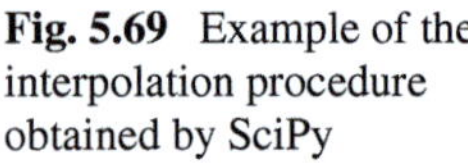

Fig. 5.69 Example of the interpolation procedure obtained by SciPy

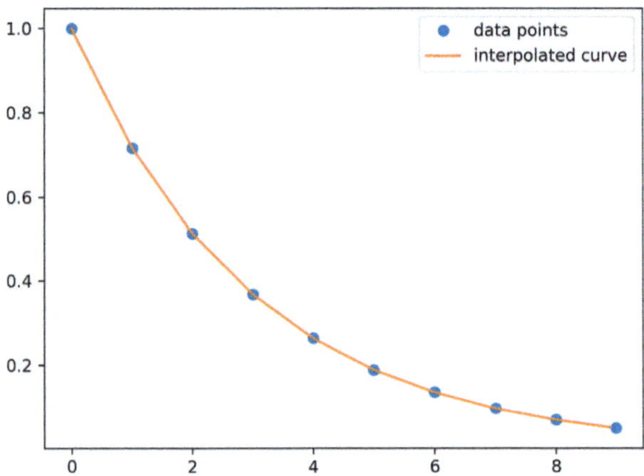


```
11
12  # New x values for interpolation
13  xnew = np.arange(0, 9, 0.1)
14  ynew = f(xnew)   # Interpolated y values
15
16  # Plotting the results
17  plt.plot(x, y, 'o', label='data points')
18  plt.plot(xnew, ynew, '-', label='interpolated curve')
19  plt.legend()
20  plt.show()
```

Code 5.150   interpolate   function in SciPy

In this example:

1. we produce some data points defined by x and y;
2. we use "interp1d" to create an interpolation function f;
3. we find interpolated values ynew for new x values xnew by means of the function f;
4. finally, we plot the original data points and the interpolated curve as illustrated in Fig. 5.69.

Providing a wide range of algorithms and functions for various tasks, we recommend learning more about the capabilities of this robust Python library through the following links:

- https://docs.scipy.org/doc/scipy/
- https://www.physics.rutgers.edu/grad/509/03_Scipy.html
- https://scipy-lectures.org/intro/.

5.6 **Data Elaboration and Visualization with PyROOT**

PyROOT is an interface provided by CERN, which allows you to use all the features of ROOT in a Python environment, providing the possibility to use ROOT classes in the analysis of extensive data, combined with the flexibility and simplicity of the Python language. Let's give a simple example of how it is convenient to use. In the Code 5.151, we implement a Python script able to read a data file containing measurements sorted in an unknown number of columns and rows, to compute the histograms of the data of each column, and to save the results in a root file.

```python
import numpy as np
import ROOT as rt
data = np.loadtxt("data10.dat")  # Load everything into a
    matrix
histograms = []  # A list that will contain the TH1F
    histograms
bins = 100
for column_index in range(len(data[0, :])):  # Loop over the
    columns
    current_column = data[:, column_index]
    histograms.append(rt.TH1F("histo" + str(column_index), "",
     bins, current_column.min(), current_column.max()))
    for value in current_column:
        histograms[column_index].Fill(value)

save_file = rt.TFile("histoall.root", "recreate")  # The '
    recreate' option
for column_index in range(len(data[0, :])):
    histograms[column_index].Write()
save_file.Close()  # If you don't close it manually, it will
    close automatically at the end of the execution
for i in range(len(histograms)):
    write_file = open("histo" + str(i) + ".dat", "w")
    for j in range(bins):
        write_file.write(str(round(histograms[i].GetBinCenter(
        j), 3)) + "   " + str(histograms[i].GetBinContent(j)) + "\n
        ")
    write_file.close()
```

Code 5.151 Histogram computation and storage example by using PyROOT

We invite the reader to rewrite the code in C++, noting how it becomes longer and more complex. The complexity of the operations remains in the PyROOT code, but Python simplifies this complexity thanks to its flexibility.

In the next exercise, we analyze data stored in a file, computing the histogram and fitting it with the proper functions. We report the final result in the Code 5.152, but the analysis proceeds need to compute and plot the histogram and to choose the best function to be used by checking its trend, as described in the previous sections.

```python
import ROOT as rt
import numpy as np
#histoname = input("give me the file name\n")
data = np.loadtxt("notsoeasy.dat")
histo = rt.TH1F("histo", "MyFit", 100, 0, 0)  # 0,0: the
    boundaries will be automatically set
for a in data:
    histo.Fill(a)
histo.Draw()
function = rt.TF1("function", "[2]+[1]*x+[0]*pow(x,2)", 3.1,
    3.7)
function.SetParNames("A", "B", "C")
function2 = rt.TF1("function2", "[0]*exp(-0.5*pow((x-[1])
    /[2],2))", 3.7, 4.3)
function2.SetParNames("D", "E", "F")
function2.SetParameters(1000, 4, 0.2)  # initial parameters
    for the fit
# the default fit method works on the variability domain of
    the histogram
# if we want it to work on the function's domain instead
histo.Fit("function2", "R")  # R stands for Range (interval)
    ... function's domain
histo.Fit("function", "R+")  # + to superimpose the other fit
sum_function = rt.TF1("sum_function", "function+function2", 3,
     5)
sum_function.SetLineColor(2)
histo.Fit("sum_function", "R")
```

Code 5.152 Histogram fit with PyROOT

By running the instructions of the Code 5.152 up to line 8 we can see the plot reported in Fig. 5.70. We observe a distinct peak superimposed on a background by examining the figure. We hypothesize that a Gaussian function can model the peak, while a second-order polynomial describes the background. As explained in previous sections, we first fit these functions over the intervals where each dominates. We then define a composite function as the sum of these two models and perform a fit on the entire data range using this composite function. We establish optimal starting values by inheriting the parameters from the individual fits, ensuring that the fit procedure converges to the best possible solution. The result of the PyROOT script is reported in Fig. 5.71.

We encourage the reader to transfer the parameters obtained from the composite function's best fit to the individual component functions, modify them as needed, and overlay them on the same figure. Additionally, the second-order polynomial function can subtract the background from the histogram. A more complex example of using PyROOT is presented in the educational exercise shown in code Listing 5.153. In this exercise, two Gaussian distributions are generated, each with 100000 events, centered at a mean value of 3 (with a standard deviation of 0.12) and 3.7 (with a standard deviation of 0.32), respectively. These distributions are overlaid on a parabolic background specified by the function $-x^2 + 5x$ over the interval $[0, 5]$ with 500000 events. The task involves fitting the resulting histogram to recover the original

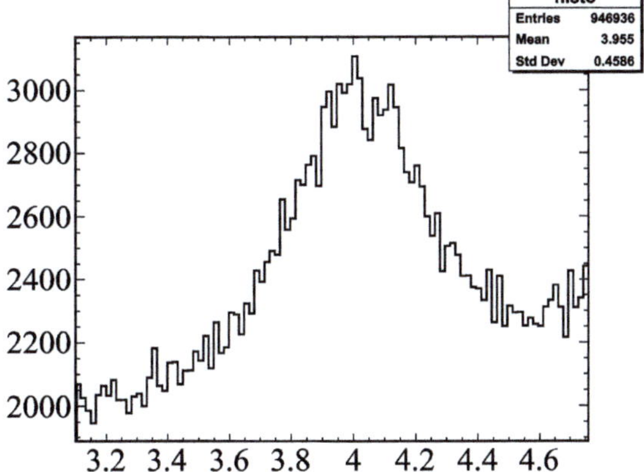

Fig. 5.70 Histogram of data stored in the file notsoeasy.dat

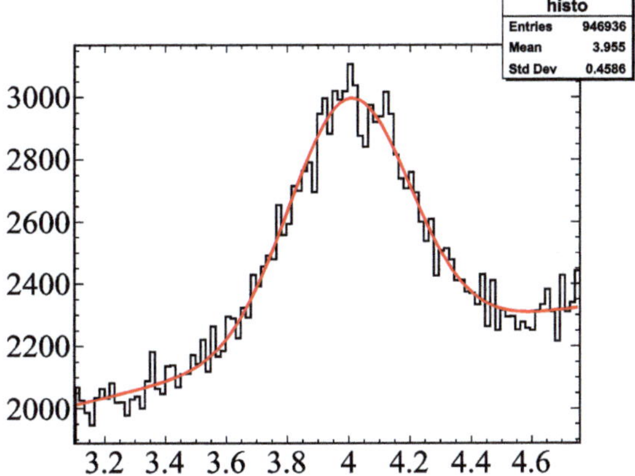

Fig. 5.71 Fitting of the histogram of data stored in the file notsoeasy.dat

parameters used in the simulation. This exercise is repeated after applying Gaussian smearing with varying standard deviations to simulate experimental resolution. The objective is to illustrate how measurement uncertainty influences the accuracy of parameter extraction from data.

```python
# 1) Generate two Gaussian distributions with 100,000 events
    each,
#    centered at and with mean values respectively: 3 and
    0.12,  3.7 and 0.32,
#    plus a parabolic background of the form -x**2 + 5*x in
    the interval 0-5 with 500,000 events.

import ROOT as rt  # type: ignore
import numpy as np

neventi1 = 100000

gaus1 = np.random.normal(3, 0.12, neventi1)
```

```python
gaus2 = np.random.normal(3.7, 0.32, neventi1)


neventi2 = 500000
par_f = rt.TF1("par_f", "-x**2 + 5*x", 0, 5)
par_random = [par_f.GetRandom() for _ in range(neventi2)]


# a) Use the generated events to create a single histogram.
c = rt.TCanvas("c", "", 1800, 1200)
h = rt.TH1F("h", "Histogram", 100, 0, 5)
for a in gaus1:
    h.Fill(a)


for b in gaus2:
    h.Fill(b)


for c in par_random:
    h.Fill(c)


h.SetLineColor(rt.kBlack)
h.Draw()


# b) Fit the histogram and assess how well the original
#    generation parameters are recovered:
#     from the code, evaluate how accurately the fit retrieves
#     the parameters used to generate the events: 3, 0.12, 3.7,
#     -1, 5, 0
g1 = rt.TF1("g1", "gaus", 2.85, 3.15)
g1.SetParNames("A", "B", "C")
g1.SetParameters(24225.4, 3.005, 0.153)
g1.Draw("same")
h.Fit("g1", "R+")


fit_result1 = h.Fit("g1", "RS")
print(f"The reduced chi2 value is {fit_result1.Chi2()/
    fit_result1.Ndf()}")


g2 = rt.TF1("g2", "gaus", 3.45, 4.10)
g2.SetParNames("D", "E", "F")
g2.SetParameters(12095.1, 3.65, 0.438)
g2.Draw("same")
h.Fit("g2", "R+")


fit_result2 = h.Fit("g2", "RS")
print(f"The reduced chi2 value is {fit_result2.Chi2()/
    fit_result2.Ndf()}")


par = rt.TF1("par", "pol2", 0, 2.65)
par.SetParNames("G", "H", "I")
par.SetParameters(6.04, 5953.81, -1169.82)
par.SetLineColor(rt.kMagenta)
par.Draw("same")
h.Fit("par", "R+")

```

```python
fit_result3 = h.Fit("par", "RS")
print(f"The reduced chi2 value is {fit_result3.Chi2()/
    fit_result3.Ndf()}")
par.SetRange(0, 5)

sum_fit = rt.TF1("sum_fit", "g1+g2+par", 0, 5)
sum_fit.SetLineColor(rt.kBlue)
sum_fit.Draw("same")
h.Fit("sum_fit", "R+")

fit_result4 = h.Fit("sum_fit", "RS")
# print(f"The reduced chi2 value is {fit_result4.Chi2()/
    fit_result4.Ndf()}")

# 1st Gaussian: mean = 3, std dev = 0.12
print(f"The parameters of the 1st Gaussian are: mean = {
    sum_fit.GetParameter('B')}, std dev = {sum_fit.
    GetParameter('C')}")
# 2nd Gaussian: mean = 3.7, std dev = 0.32
print(f"The parameters of the 2nd Gaussian are: mean = {
    sum_fit.GetParameter('E')}, std dev = {sum_fit.
    GetParameter('F')}")
# a*x**2 + b*x + c = -x**2 + 5*x
print(f"The parameters of the parabolic background are: a = {
    sum_fit.GetParameter('G')}, b = {sum_fit.GetParameter('H')
    }, c = {sum_fit.GetParameter('I')}")

print("The parameters of the two Gaussians obtained from the
    fit match the original ones.")

# c) Repeat exercise b) applying Gaussian smearing to the two
    distributions
#     with standard deviation 0.1 and 1. Comment on what
    happens in both cases.

# Create a for loop over the given sigma values.
# For each loop, create a canvas and a histogram for the
    current sigma.
# Each histogram is fitted with functions defined using
    previous fit parameters.
# For each histogram, compare the fitted parameter values.

sigmas = [0.1, 1]
canvas = []
histograms = []

for a in sigmas:
    c2 = rt.TCanvas(f"c2{a}", f"Canvas {a}", 1200, 800)
    canvas.append(c2)

    gaus1_sm = gaus1 + np.random.normal(0, a, neventi1)
    gaus2_sm = gaus2 + np.random.normal(0, a, neventi1)
```

```python
 99    h2 = rt.TH1F(f"h2{a}", f"Distributions with smearing:
       sigma = {a}", 100, 0, 7)
100    histograms.append(h2)
101
102    for i in gaus1_sm:
103        h2.Fill(i)
104    for j in gaus2_sm:
105        h2.Fill(j)
106    for k in par_random:
107        h2.Fill(k)
108        h2.SetLineColor(rt.kBlack)
109    h2.Draw()
110    c2.Update()
111
112    c2.cd()
113    g1_sm = rt.TF1("g1_sm", "gaus", 2.85, 3.15)
114    g1_sm.SetParNames("L", "M", "N")
115    g1_sm.SetParameters(24225.4, 3.005, 0.153)
116    g1_sm.Draw("same")
117    h2.Fit("g1_sm", "R+")
118
119    fit_result5 = h2.Fit("g1_sm", "RS")
120    print(f"The reduced chi2 value is {fit_result5.Chi2()/
       fit_result5.Ndf()}")
121
122    g2_sm = rt.TF1("g2_sm", "gaus", 3.45, 4.10)
123    g2_sm.SetParNames("O", "P", "Q")
124    g2_sm.SetParameters(12095.1, 3.65, 0.438)
125    g2_sm.Draw("same")
126    h2.Fit("g2_sm", "R+")
127
128    fit_result6 = h2.Fit("g2_sm", "RS")
129    print(f"The reduced chi2 value is {fit_result6.Chi2()/
       fit_result6.Ndf()}")
130
131    par_sm = rt.TF1("par_sm", "pol2", 0, 2.65)
132    par_sm.SetParNames("R", "S", "T")
133    par_sm.SetParameters(6.04, 5953.81, -1169.82)
134    par_sm.SetLineColor(rt.kMagenta)
135    par_sm.Draw("same")
136    h2.Fit("par_sm", "R+")
137
138    fit_result7 = h2.Fit("par_sm", "RS")
139    print(f"The reduced chi2 value is {fit_result7.Chi2()/
       fit_result7.Ndf()}")
140    par_sm.SetRange(0, 5)
141
142    sum_sm = rt.TF1("sum_sm", "g1_sm + g2_sm + par_sm", 0, 7)
143    sum_sm.SetLineColor(rt.kBlue)
144    sum_sm.Draw("same")
145    h2.Fit("sum_sm", "R+")
146
147    fit_result8 = h2.Fit("sum_sm", "RS")
```

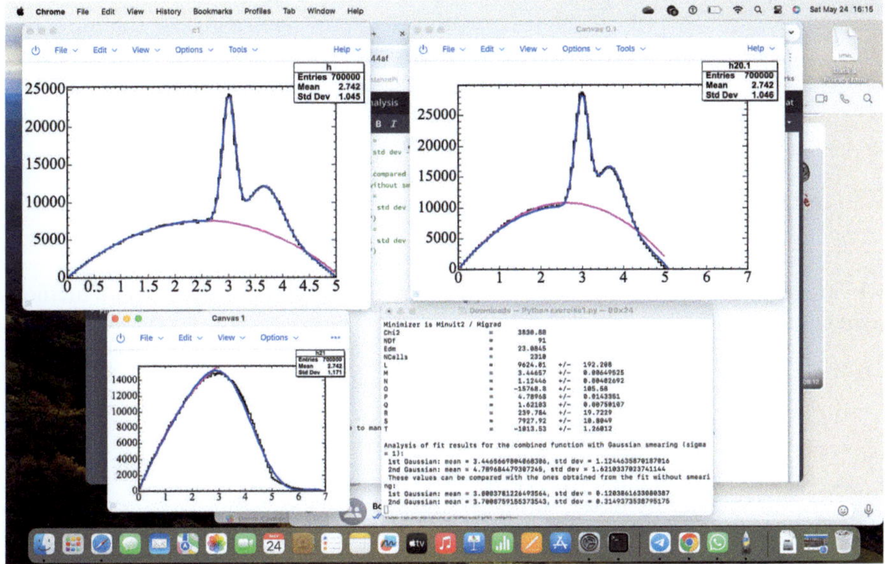

Fig. 5.72   Results of the running Code 5.153

```
148    # print(f"The reduced chi2 value is {fit_result8.Chi2()/
       fit_result8.Ndf()}")
149    c2.Update()
150
151    print(f"\nAnalysis of fit results for the combined
       function with Gaussian smearing (sigma = {a}):")
152    print(f" 1st Gaussian: mean = {sum_sm.GetParameter('M')},
       std dev = {sum_sm.GetParameter('N')}")
153    print(f" 2nd Gaussian: mean = {sum_sm.GetParameter('P')},
       std dev = {sum_sm.GetParameter('Q')}")
154    print(" These values can be compared with the ones
       obtained from the fit without smearing:")
155    print(f" 1st Gaussian: mean = {sum_fit.GetParameter('B')},
        std dev = {sum_fit.GetParameter('C')}")
156    print(f" 2nd Gaussian: mean = {sum_fit.GetParameter('E')},
        std dev = {sum_fit.GetParameter('F')}")
157
158 input("")
```

Code 5.153   A scholar exercise soved with pyroot

Figure 5.72 presents the results of executing the script shown in code Listing 5.153. The figure displays the fits performed in the three scenarios, while the terminal output reports all the statistical quantities estimated during the fitting procedure.

5.7   Exercises

5.1 Make the plot of the following function in the interval $(-10, 10)$ using all methods learned in this chapter:

$$f(x) = \begin{cases} \sin(x)/x & \text{se } x \leq 0 \\ -\sin(x)/x & \text{se } x > 0 \end{cases}$$

and save the plot in a file in PDF format.

5.2 Write a macro that plots the functions $f(x) = x^2$ and $g(x) = 5 - x^2$ over the interval $[-10, 10]$. Also plot the sum function $f(x) + g(x)$ and the difference function $f(x) - g(x)$.

What do the points where the difference function intersects the x-axis represent with respect to the two component functions?

5.3 Estimate the number of signal events in the histogram using the data from the file newfittolino.dat. Use the file newplottamo.dat to create three plots using the TGraph class, using the first column as the x-axis for all plots, and the other columns as the y-axis. Save the plots in a TFile.

5.4 Write a code representing the distribution of experimental data provided in Table 5.5. Perform a best fit of the data using an appropriate model. Save the fit parameters, their associated errors, and the reduced chi-squared value to a file.

Table 5.5 Measurement data with corresponding uncertainties

Angle (°)	Period (s)	Errors on angle (°)	Errors on period (s)
90	9.630	0.1	0.0352134
80	9.496	0.1	0.0392936
70	9.364	0.1	0.0492339
60	9.286	0.1	0.0101980
50	9.384	0.1	0.0361108
40	9.564	0.1	0.0449888
30	9.620	0.1	0.0603322
20	9.838	0.1	0.0312410
13	9.996	0.1	0.0332267

5.5 Write the function $f(x) = \frac{\sin(x)}{x^2}$ in the interval $[-5, 5]$. Plot the graph. Print to the screen the values of f(x) corresponding to a numerical sequence from -5 to 5 with a step of 0.01.

Chapter 6
Experimental Data Acquisition and Analysis

6.1 Introduction

Natural sciences are based on theory, valid experimental evidence, criticism, and rational discussion. Experiments play a fundamental role because they allow us to test hypotheses and/or create new ones, furnishing the basis for scientific knowledge. In particular, physics uses experiments to investigate the properties of matter and the processes/phenomena that occur within the universe by studying essential quantities such as motion and energy. In this frame, the combination of physics and computer science permits to design and realize low-cost experiments to (i) better understand problems, (ii) explore several processes, even simulating systems and testing the limits of a particular model, and (iii) improve data acquisition, the analysis and the visualization of the results. In this chapter, we show to the reader how all of this can be done with using the open source software already seen in the previous chapters.

6.2 Microcomputers and Microcontrollers

Microcomputers or Single-Board Computer SBC can be identified as small-sized and low cost computing devices with a microprocessor as their central processing unit (CPU). They have installed a single printed circuit board with a microprocessor, memory, and input/output (I/O) circuitry. We can say that many microcomputers are also personal computers when equipped with a keyboard and screen for input and output. In fact, due to their lower costs, individuals can possess them as their personal computers. At first, microcomputers were not very powerful. Just keep in mind that the Commodore 64—one of the first and the most popular microcomputers of its era and is the best-selling model of home computer of all time—included, for example, a microprocessor with a clock frequency of 1 MHz, Read Only Memory (ROM) of 20 kB and Random Access Memory (RAM) of 64 kB: they were very high values for

© The Author(s), under exclusive license to Springer Nature Switzerland AG 2026
S. Vasi et al., *Open Source Tools for Physics Data Analysis*, Undergraduate Texts
in Physics, https://doi.org/10.1007/978-3-032-00721-6_6

these components at that time, but very ... very low compared to a microcomputers employed today. In Sect. 6.2.1 we will see one of the most popular microcomputer used today, the Raspberry Pi.

Concerning *microcontrollers*, they are small and inexpensive electronic devices integrated on an electronic circuit, designed to perform only specific tasks for specific embedded system applications. Generally speaking, a microcontroller consists of the processor, the memory (RAM, ROM, EPROM), Serial ports, peripherals (timers, counters), etc., but it does not have an OS and it has far fewer resources than SBCs. For instance, microcontrollers may have KBs of storage instead of MBs or GBs of SBCs. Their processing capabilities are limited, making them more *task oriented* than microcomputers—a more general-purpose computing device. In this frame, a microcontroller can be employed in various applications such as light and temperature sensing, and controlling devices like LED, and/or measuring devices like a Volt Meter. Among the programmable boards with microcontrollers currently used must indeed be cited Arduino, which will be described within the Sect. 6.2.2.

6.2.1 Raspberry Pi

Raspberry Pi (RPi) is a single-board computer designed to host operating systems based on the Linux kernel—its specially dedicated operating system is called Raspberry Pi OS, but it is possible to use different operating system distributions (e.g., Kali, Linux Pidora, etc.)—that has obtained an extraordinary success worldwide for its tiny dimensions, its power and integrated components, as well as its low price. There are several RPi models, but all of these are based on a Broadcom-manufactured system-on-a-chip, which incorporates an ARM processor, a VideoCore IV GPU, and an amount of memory depending on the model. The RPis do not include hard disks or solid state drives; instead, they rely on an SD card for booting and non-volatile memory. RPis provides a Camera Pinout—a dedicated MIPI CSI standard (Camera Serial Interface) port—a set of GPIO (General Purpose Input/Output) pins and an Inter Integrated Circuit (I^2C)—a two-wire serial communication system used between integrated circuits—allowing you to control electronic components for physical computing and explore the Internet of Things (IoT). One of the most essential features of RPi is the presence of the 40-pin GPIO header—for the Raspberry Pi 4. This header provides two 5 V outputs, two 3, 3 V, 8 ground pins, and many GPIO pins devoted to bus communications, pulse width modulation (PWM) pin working in input or output mode.

In particular, the importance of using these pins will be evident when discussing sensors.

These features have allowed the Raspberry Pi to be used for many purposes. People worldwide use RPi to learn programming skills, build hardware projects, and even use it in industrial applications.

Below is a summary list of the most recent and/or used RPi models:

- Pi 3 Model B+ (2018)
- Pi 4 Model B (2020)
- Pi Zero 2 W (2021)
- Pi 5 (2023).

A different market price characterizes each of these models depending on the hardware they are made of and, therefore, the performance, but, generally speaking, every RPi allows realizing cheap and portable experiments involving STEM topics such as physics, mathematics, and computer science. As a starting point for an experiment, whatever RPi model is chosen, it is necessary to carry out an initial setup of the single-board computer. Note that here we consider the reference model the Raspberry Pi 4 Model B because it is the most chosen one by the people, and it's generally easier to start a project with RPi 4, then move to Raspberry Pi Zero if possible. First of all, we have to power up the machine and install the Linux software—it takes about an hour to be ready—making use of:

- a Raspberry Pi,
- a power supply,
- a microSD card,
- a keyboard, a mouse, an HDMI cable and a monitor.

The initial configuration of our RPi is beyond the scope of this book. Still, we provide a handy and comprehensive guide—furnished by the Raspberry Pi Foundation at https://projects.raspberrypi.org/en/projects/raspberry-pi-setting-up/—to make it easier for the reader.

6.2.2 Arduino

Arduino is an Italian open-source electronics platform based on easy-to-use hardware and software for building digital projects and devices. Arduino board uses various microprocessors and controllers and includes digital and analog input/output (I/O) pins that may be connected to various expansion boards (SHIELDS), breadboard, and other circuits. The boards are powered with a AC-to-DC adapter or battery. Furthermore, the boards have serial communications interfaces, such as Universal Serial Bus (USB), that are also exploited to load software programmed by the user. In this frame, it is possible to write codes for the microcontroller with the C and C++ programming languages (Embedded C) using a standard Application Programming Interface (API)—the Arduino Programming Language—and the open-source Arduino Software. The Arduino project provides an Integrated Development Environment (IDE) and a command line tool useful for writing and testing code using a serial monitor for debugging and real-time data monitoring. The IDE provides a simple and intuitive code editor with syntax highlighting and auto-completion, allowing you to easily compile and upload sketches (programs) to Arduino boards. Furthermore, the IDE includes a library manager that makes it easy to install and update libraries needed for your projects. It supports a wide range of Arduino boards and can be used with various operating systems, such as Windows, macOS, Linux, and Chromebooks.

Arduino boards are many and different in both price and performance, but certainly the most popular among those taking their first steps in creating circuits is the Arduino UNO. It has 20 separate channels for INPUT and OUTPUT: all these channels can be used to read digital signals (zeros and ones), and 6 are dual-use and can also read analog signals. In addition, Arduino UNO is the best choice because most of the shields and the additional boards and accessories have been created for this version of Arduino, and because you will find a lot of applications/codes already available, provided by the community—if you are looking for ready-to-go software.

A very brief discussion on SHIELDs should be made. Each of them contains many accessory parts already incorporated into the board, allowing developers to focus on software and algorithm development and save the time they would waste buying, connecting, and testing additional cards.

As already seen for the RPi in Sect. 6.2.1, Arduino boards can also be used to realize cheap and portable STEM experiments. We will show you some simple experiments realized by programming the Arduino board in C++, comparing what was obtained with Python and an RPi.

In the next section, we will introduce sensors, which are used to measure physical quantities and carry out real scientific experiments.

6.3   Sensors

Sensors produce an output signal to sense and measure a physical phenomenon. The input signal can be heat, pressure, moisture, light, etc., instead, the output is usually a digital or analog human-readable signal—possibly another more easily readable and comparable physical quantity. They have a role in the Internet of Things (IoT), making it important for them to realize systems for monitoring, managing, and controlling a precise physical environment more easily and efficiently. IoT sensors are used in everyday life and countless applications that most people are unaware of, e.g., temperature measurements or weather stations. Using sensors with the Raspberry Pi or an Arduino is a great way to extend the board's functionality without considerable effort or expenses. If sensors are linked with a microcomputer, they give the possibility not only to acquire data, but also to record and save it for further processing if programmed efficiently with our Open-Source code.

There are several ways to classify sensors, but we commonly categorize them (i) depending on the physical quantity being measured, (ii) depending on whether they are passive or active, and (iii) depending on whether they are analog or digital. Concerning the environmental factors that the sensors measure, for example, we can distinguish:

- **Temperature sensors**, identifying the temperature of a target medium, whether gas, liquid or air.
- **Pressure sensors**, evaluating the pressure of a liquid or gas, and, eventually, monitoring the flow of gases or liquids.
- **Humidity sensors**, determining the relative humidity using the level of water vapors (moisture) in the air. Note that these sensors often comprehend temperature readings because relative humidity depends on the air temperature.
- **Light sensors**, also called photosensors, detect light in ultraviolet, visible, and infrared range.
- **Accelerometers** measuring acceleration, tilt, and vibration.
- **Motion and proximity sensors**, evaluating the presence and the physical movement in a defined space of an object, or determining the distance between objects.
- **Force sensors**, force transducers converting an input mechanical force, such as load, weight, tension, compression, or pressure, into another physical variable, such as an electrical output signal.
- **Magnetometers**, measuring the intensity of magnetic fields and sensing where the strongest magnetic force is coming from.

The above-mentioned sensors are only a few of the several types of sensors being used for evaluating physical quantities across environments and within devices.

We can distinguish passive or active sensors depending on the interaction with the environment in which the sensor is deployed. Passive sensors work by exploiting the sources of the environment in which they are placed to make a measurement, e.g., thermal energy. By contrast, an active sensor needs an external power source to generate an output from an input of the physical environment in which the sensor is placed. Anyway, some sensors exist in both passive and active forms, e.g., infrared light sensors.

Finally, according to their output, sensors can be analog or digital. Analog sensors generate ongoing and changing output analog signals from the input. Thermocouples used in gas hot water heaters offer a good example of analog sensors. The water heater's pilot light continuously heats the thermocouple. If the pilot light goes out, the thermocouple cools, sending a different analog signal that indicates the gas should be shut off. On the other hand, digital sensors produce discrete digital signals as binary numbers (1s and 0s). Like active and passive sensors, some sensors are available in both analog and digital forms. In this case, too, the environment in which the sensor will operate typically determines which is the best option.

Here is a list of some sensors that can be used to realize experiments evaluating physical quantities.

In the next sections, we will see some examples of how to program these sensors using Python or C++ for Arduino and Raspberry.

Table 6.1 Physical quantities and list of sensors

Sensor	Used for evaluate
DHT11, DS18B20, DS18S20	Temperature
DHT22	Humidity and temperature
VEML6075	UV light index
GM5516	Light intensity
BMP180	Atmospheric pressure, temperature and elevation
KY-035	Magnetic field
Force sensitive resistor (FSR)	Force, pressure or mechanical stress

6.4   Experiments

As already stated previously, also in the introduction—Sect. 6.1—to this chapter, experiments are fundamental to the growth of science. They allow us to obtain results to explain a physical phenomenon. Still, often there are variables—such as environmental conditions—that can lead to different outcomes in experimentation. It is obvious how critical it is to have an experiment that should be repeatable under similar conditions to allow scientists to validate a theory and make a measurement trustworthy.

All the open-source tools seen in this book, coupled with the elements shown in this chapter, enable us to perform physics experiments and analyze the data obtained from them. In this frame, in the following subsections, we show several examples for measuring physical quantities using microcomputers and/or microcontrollers coupled with sensors, and then the codes used to analyze the data obtained from the measurement procedure.

6.4.1   DS18B20 Sensor Temperature Monitoring

One of the most basic, but valuable and typical experiments is the one related to the measurement of *temperature*. The physical quantity expresses the attribute of hotness or coldness in several arbitrary scales, e.g., the Kelvin (K) scale, an absolute temperature scale recognized as the international scientific standard. This kind of measurement is so crucial that we unconsciously evaluate an object's temperature at least once every day without giving it too much thought. For example, each smartphone measures the temperature of its processor using a sensor that acts as a thermometer.

As shown in the Table 6.1 reported in Sect. 6.3, we can use several sensors to measure the temperature of a physical system. For example, DS18B20 is a digital sensor we can use. It is a waterproof one-wire device with a unique 64-bit serial code, allowing us to wire multiple sensors to the same data wire. Specifically, it can be powered with a voltage in a range of 3.0–5.5 V and it measures temperatures from $-55\,°C$ to $+125\,°C$ with an accuracy of $\pm0.5\,°C$. Let's see how to create a temperature measurement experience with subsequent data collection and analysis using Python or C++ to control our sensor connected with a Raspberry Pi or Arduino.

6.4.1.1   Raspberry Pi and Python

Before starting, here is a list of the objects necessary to carry out the experience:

- Raspberry Pi Board with a micro SD card (with USB keyboard and mouse)
- DS18B20 sensor
- $4.7k\Omega$ resistor
- Jumper wires
- Breadboard.

In creating our experimental setup, we must first know our sensor as best we can to understand how to connect it correctly to our Raspberry.

Tip
The best way to have all the information regarding our sensor is its *datasheet* that we can usually find in e-commerce sites that sell the product.

The waterproof version of the DS18B20 sensor consists of a sealed digital temperature probe—which gives the possibility to measure temperatures in wet environments —with an onboard analog-to-digital converter (ADC) and a simple pinout to easily hook it up to the Pi to power the sensor and acquire data from it. Specifically, the waterproof version is usually characterized by the following pinout:

- Red wire = Voltage at the Common Collector (VCC), the positive supply voltage;
- Black wire = Ground (GND), connected to the negative terminal of a power supply;
- Yellow wire = DATA, output cable connected to a digital pin of our Raspberry Pi.

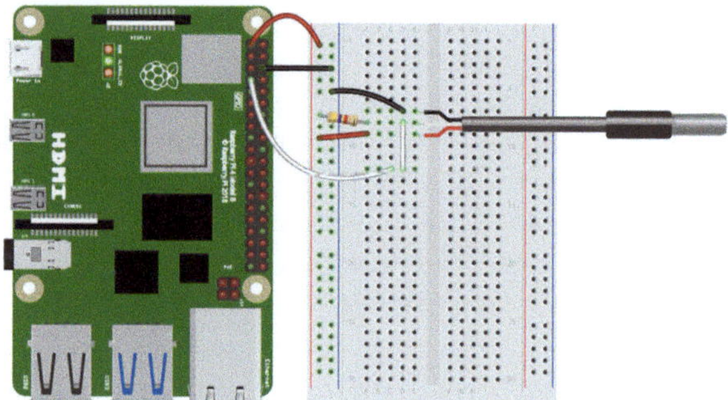

Fig. 6.1 Setup for temperature measurements with Raspberry Pi and DS18B20 sensor. This image was created with Fritzing

The connection is simple, as indicated in Fig. 6.1 created with Fritzing. First of all, we have to connect (i) the red cable to 5 or 3.3 V, (ii) the black wire to GND and (iii) the DATA signal cable (the white one) to a digital pin. In the example, the latter is connected to GPIO #4 (Pin #7 in Raspberry Pi if using physical numbering) with a 4.7 K Ohm resistor between the signal and the power pin (5 or 3.3 V).

One-Wire Interface

The One-Wire interface must be enabled before the Pi can receive data from the DS18B20 sensor. The default 1-Wire GPIO for the Raspberry Pi is GPIO #4. So, once the connection is all settled, power up your Pi and log in, and it is possible to enable the one-wire interface in two ways.

1. Type the following command via terminal and press Enter:

```
sudo raspi-config
```

   Figure 6.2 shows the new window that will open from which you can select "Interface Options": Pressing Enter, you can now select the "1-Wire" option, as showed in Fig. 6.3. Now you can enable the One-Wire interface clicking on "Yes", selecting "Finish" (Fig. 6.4).
2. We can make use of the "Raspberry Pi Configuration" tool, selecting it from the "Preferences" menu: When it opens, click on the "Interfaces" tab and then click on "Enable" for the "1-Wire interface", as illustrated in Fig. 6.5. Then, press "OK".

   Whichever method you choose to enable the 1-Wire configuration, restart your raspberry Pi.

The following script prints temperature readings in the shell from the DS18B20 temperature sensor. Create a new Python file (for example via the Python 3 Editor— IDLE) and copy the following code.

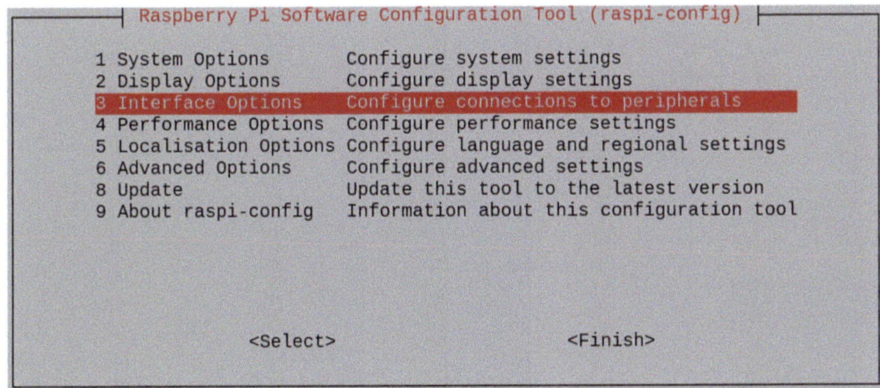

Fig. 6.2 Inside the "Raspberry Pi Software Configuration Tool" (raspi-config), select the "Interface Options" choice

Fig. 6.3 "1-Wire" selection

```python
import os
import glob
import time

os.system('modprobe w1-gpio')
os.system('modprobe w1-therm')

b_dir = '/sys/bus/w1/devices/'
dev_folder = glob.glob(b_dir + '28*')[0]
dev_file = dev_folder + '/w1_slave'

def read_t_raw():
    file = open(dev_file, 'r')
    lines = file.readlines()
    file.close()
```

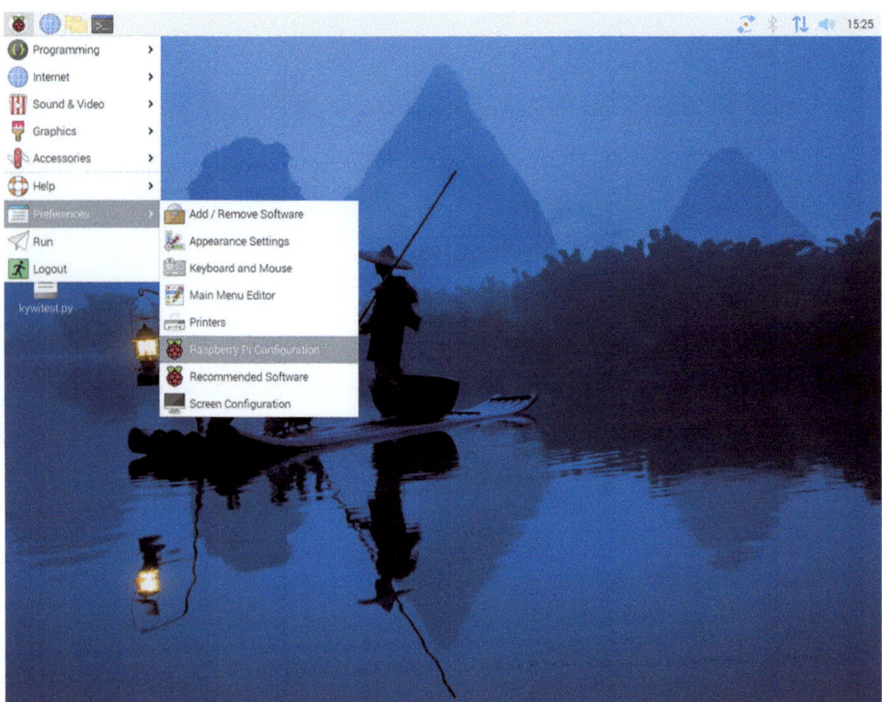

Fig. 6.4 Screenshot Raspberry Pi OS Desktop showing how to select the "Configure Rasberry Py System"

```python
16         return lines
17
18 def read_t():
19     lines = read_t_raw()
20     while lines[0].strip()[-3:] != 'YES':
21         time.sleep(0.2)
22         lines = read_t_raw()
23     equals_pos = lines[1].find('t=')
24     if equals_pos != -1:
25         t_string = lines[1][equals_pos+2:]
26         t_c = float(t_string) / 1000.0
27         return t_c
28
29 while True:
30     data=(read_t())
31     print data, 'C'
32     time.sleep(1)
```

Code 6.1 Raspberry Pi, Python and DS18B20 sensor

On the other hand, it would also be possible not only to see the values acquired by the sensor on the screen but also to record the data by adding such strings of code:

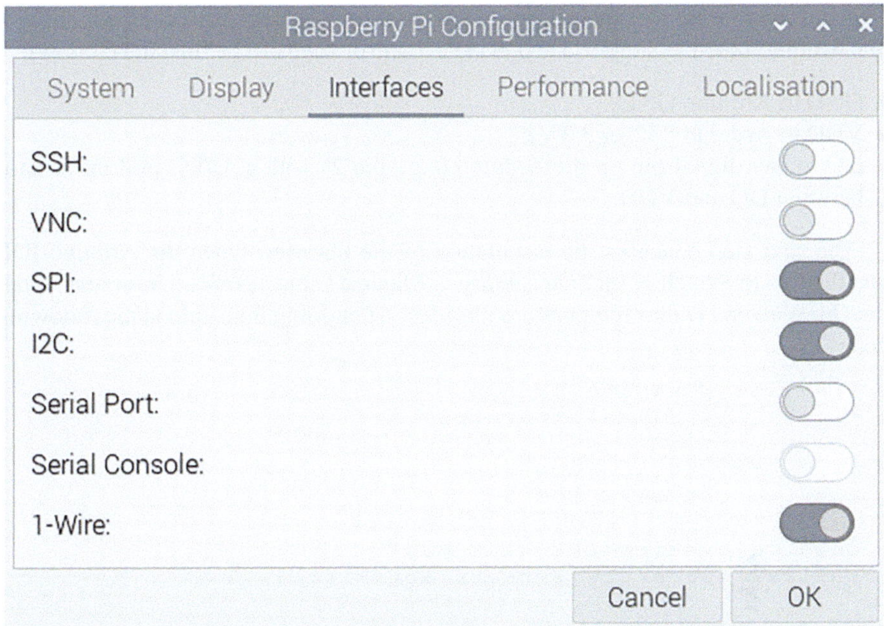

Fig. 6.5 Click on "Enable" for the "1-Wire interface" inside the "Interfaces" tab

```
1  ...
2  while True:
3      data=(read_t())
4      with open("/var/tmp/temp.txt","a") as file:
5          file.write(data + "\n")
6      print data, 'C'
7      time.sleep(1)
```

The "with open()" command makes the file shown in parentheses usable and closes it after running the next lines of code. In this way, upon closing the script we are able to reuse the data file for later analysis and visualization.

6.4.1.2 Arduino and C++

We can also reproduce the same experiment by using an Arduino programmed with C++. In this case, we need the following objects to realize the experience:

- Arduino board (e.g., Arduino UNO)
- DS18B20 sensor
- 4.7kΩ resistor
- Jumper wires
- Breadboard.

Similarly to what we saw in the previous paragraph, wiring these components with our Arduino board is simple. The DS18B20 sensor needs to be linked as follows:

- GND to Arduino GND;
- VDD to Arduino 5 V (or 3.3 V);
- DQ to any digital pin on the Arduino (e.g., pin 2) with a 4.7kΩ pull-up resistor between DQ and VDD.

The next step concerns the installation of the libraries. Open the Arduino IDE and then go to Sketch > Include Library > Manage Libraries. Search for and install the OneWire and DallasTemperature libraries. After doing this, upload the following code:

```cpp
#include <OneWire.h>
#include <DallasTemperature.h>

// Data wire is connected to pin 2
#define ONE_WIRE_BUS 2

OneWire oneWire(ONE_WIRE_BUS);
DallasTemperature sensors(&oneWire);

void setup() {
  Serial.begin(9600);
  sensors.begin();
}

void loop() {
  sensors.requestTemperatures();
  float temperatureC = sensors.getTempCByIndex(0);
  Serial.print("Temperature: ");
  Serial.print(temperatureC);
  Serial.println(" C");
  delay(1000);
}
```

Code 6.2 Arduino, C++ and DS18B20 sensor

6.4.2 Enviro+ Environment Monitoring

The Enviro+ board is designed to carry out various experiments on physics with a focus on environmental monitoring, both indoors and outdoors. It consists in several sensors allowing the measurement of the concentration of pollutant gases and particulates, temperature, pressure, humidity, light, and noise levels. A list of the hardware installed on the board is given below:

- BME280 sensor: Measures temperature, pressure, and humidity.
- LTR-559 sensor: Measures light and proximity.

- MICS6814 sensor: Detects various gases including carbon monoxide, nitrogen dioxide, and ammonia.
- MEMS microphone: Measures noise levels.
- "0.96" color LCD: Displays the collected data.
- Matter sensor: Allows for additional air quality monitoring.

A Python library to control all the parts of the Enviro+ board is already furnished by the producers along with guides and examples to install and use the card (see this link).

Here, we present a simple example of making use of this board with a Raspberry Pi Zero WH[1] to measure some of the previous physical quantities reported:

```python
#!/usr/bin/env python3

import time
import csv
import logging
import datetime
from bme280 import BME280

# Import the SMBus module, using smbus2 if available
    , otherwise smbus
try:
    from smbus2 import SMBus
except ImportError:
    from smbus import SMBus

# Configure logging to record information with a
    timestamp
logging.basicConfig(
    format='%(asctime)s.%(msecs)03d %(levelname)-8s
    %(message)s',
    level=logging.INFO,
    datefmt='%Y-%m-%d %H:%M:%S')

# Startup message for the program
logging.info("""weather.py - Print readings from the
    BME280 weather sensor.

Press Ctrl+C to exit!

""")

# Initialize the I2C bus
bus = SMBus(1)
# Initialize the BME280 sensor
bme280 = BME280(i2c_dev=bus)
```

---

[1] The Raspberry Pi Zero WH is a compact, low-cost computer that comes with built-in WiFi and Bluetooth, and features a pre-soldered GPIO header for easy connection to breadboards and accessories12. This makes it ideal for various projects, especially for those who prefer not to do the soldering themselves.

```python
32
33 # Create a data file with the current timestamp in
      the name
34 name_datafile = 'data_' + str(datetime.datetime.now
      ()) + '.txt'
35
36 # Create and immediately close the file to ensure it
      exists
37 datafile = open(name_datafile, "w")
38 datafile.close()
39
40 # Reopen the file in append mode
41 datafile = open(name_datafile, "a")
42
43 # Write the field names to the file
44 field_names = ['Date;', 'Temperature ( C);', '
      Pressure (hPa);', 'Humidity (%);']
45 datafile.writelines(field_names)
46
47 # Infinite loop to read and record data from the
      sensor
48 while True:
49     # Get the current timestamp
50     x = datetime.datetime.now()
51     # Read the temperature from the sensor
52     temperature = bme280.get_temperature()
53     # Read the pressure from the sensor
54     pressure = bme280.get_pressure()
55     # Read the humidity from the sensor
56     humidity = bme280.get_humidity()
57     # Log the data
58     logging.info("""\n Temperature: {:05.2f} *C \n
      Pressure: {:05.2f} hPa \n Relative humidity:
      {:05.2f} % \n """.format(temperature, pressure,
      humidity))
59     # Write the data to the file
60     datafile.writelines('\n')
61     datafile.writelines([str(x), ';', str(
      temperature), ';', str(pressure), ';', str(
      humidity)])
62     # Ensure the data is written immediately
63     datafile.flush()
64     # Wait 1 second before repeating the loop (
      modifiable to 1800 seconds to record data every
      30 minutes)
65     time.sleep(1) # value in seconds, 1800 to record
      data every 30 minutes
```

Code 6.3 Enviro+ and Raspberry Pi Zero WH

This Python script reads data from the BME280 sensor (temperature, pressure, and humidity) and logs it to a text file with a timestamp. It uses the `logging` module to print the data to the terminal and the `time` module to create an interval between readings. It is possible to adjust the time interval to suit your needs.

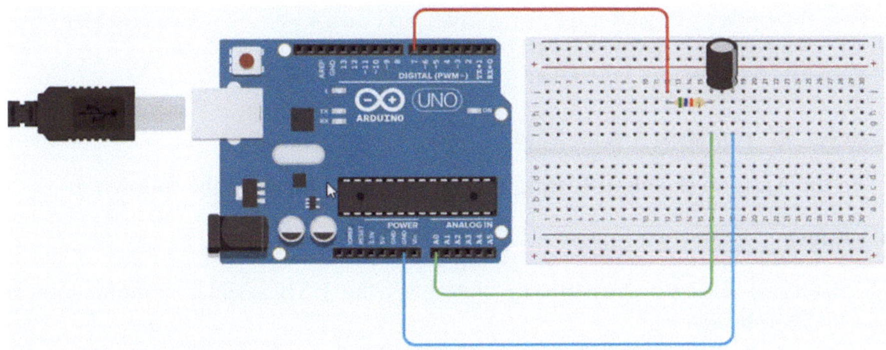

Fig. 6.6 Scheme charge-discharge RC experiment with Arduino Uno

6.4.3 Charge and Discharge of RC Circuits

A classic hands-on exercise with just an Arduino Uno, a breadboard, one capacitor, and one resistor is the charge–discharge test of an RC circuit. This experiment, a staple of every introductory physics lab, bridges theory and practice: you assemble a real circuit, capture live data, and verify that the measured time constant τ matches the theoretical value $\tau = RC$. Running it on an Arduino replaces the bulky bench oscilloscope with code you can write and execute at home. A digital pin becomes a 0–5 V square-wave generator that repeatedly charges and discharges the capacitor, while an analog pin samples the capacitor voltage so you can plot V(t), fit the exponential curve, and extract τ. The wiring diagram is shown in Fig. 6.6. Programmatically, you only need to set the digital pin HIGH and LOW with a chosen period, and read analogRead() at a suitable sampling rate, storing the values for later analysis.

To perform the experiment and to acquire the data, one can write the following C code and then load the code in the microcontroller using the Arduino IDE:

```
int ledPin = 7;          // select the pin for the LED
float event_time=0.;
float tempone=0;
int    spione =1;
// the setup routine runs once when you press reset:
void setup() {
  // initialize serial communication at 9600 bits
     per second:
  Serial.begin(9600);
  pinMode(ledPin, OUTPUT);
}
// the loop routine runs over and over again forever
   :
void loop() {
   event_time=millis();
   if(event_time>tempone && spione==1)   {digitalWrite
      (ledPin, HIGH); spione=2;tempone+=500;}
```

```
15   if(event_time>tempone && spione==2) { digitalWrite
       (ledPin, LOW); spione =1;tempone+=500;}
16   Serial.println(analogRead(A0)*5/1023.);
17 }
```

Code 6.4 RC experiment Arduino Scratch

In the Arduino sketch of Listing 6.4, the two key blocks do the following:

- **setup**():

  - `Serial.begin(115200);` initializes the USB serial communication at 115200 baud.
  - `pinMode(7, OUTPUT);` sets digital pin 7 as an output to drive the RC circuit.

- **loop**():

  - Line 13: Captures the timestamp using `millis()`, the internal millisecond counter.
  - Lines 14–15: Toggles pin 7 `HIGH` and then `LOW`, creating a $0 \rightarrow 5$ V square pulse, with its period controlled via `delay()` or `delayMicroseconds()`.
  - Line 16: Reads the capacitor voltage using `analogRead()`, converts the ADC value to volts, and sends it over the serial port.

Opening the Serial Plotter in the Arduino IDE provides a real-time visualization of the charge/discharge curve, offering a simple oscilloscope-like view. While useful for quick validation, it is less ideal for thorough data logging, as the IDE does not timestamp or store values efficiently.

Another approach involves uploading the Firmata sketch to the Arduino board using the Arduino IDE. This firmware allows the board to be controlled via Python. Once Firmata is installed on the board, the `pyfirmata` library must be installed to enable Python to interface with Arduino. This library facilitates direct communication and control of the board without requiring additional code modifications. The Python script presented in Listing 6.5 replicates the core functionality implemented in Listing 6.4. Additionally, leveraging the `matplotlib` and `NumPy` libraries enables real-time visualization of the acquired data and the ability to store measurements in a file for further analysis.

```
1  import pyfirmata
2  import time
3  import matplotlib.pyplot as plt
4  import numpy as np
5  plt.ion()
6  plt.show()
7  t = []
8  v = []
9  board = pyfirmata.Arduino('/dev/ttyACM0')#Arduino
       orginale
10 #board = pyfirmata.Arduino('/dev/ttyUSB0')
```

```python
it = pyfirmata.util.Iterator(board)
it.start()
analog_input = board.get_pin('a:0:i')
led = board.get_pin('d:7:o')
previous = time.perf_counter()
tempo = 0.
on = 1
off = 1
measure = 0.0

while tempo <4:
    now = time.perf_counter()
    deltat = now - previous
    previous = now
    tempo = tempo + deltat
    time.sleep(0.005)
    measure = analog_input.read()
    t.append(tempo)
    v.append(0 if measure is None else measure*5)
    plt.plot(t,v)
    plt.pause(0.001)
    plt.draw()
    print(tempo,"   ",measure)
    if tempo > 0 and on == 1:
        led.write(1)
        on = 0
    if tempo > 2 and off == 1:
        led.write(0)
        off = 0

times = np.array(t)
volts = np.array(v)
outputFile='RCSeguenza.dat'
np.set_printoptions(precision = 20)
np.savetxt(outputFile, np.column_stack((times, volts
    )))
```

Code 6.5 RC experiment Arduino Scratch powered by matplotlib and NumPy

To visualize the results in ROOT style, we can implement a straightforward macro as described in Listing 6.6.

```cpp
{
        TCanvas pina;
TGraph pino("RCSeguenza2.dat");
pino.GetXaxis()->SetNdivisions(905);
pino.GetXaxis()->SetTitle("time (s)");
pino.Draw();
pino.GetYaxis()->SetTitleOffset(0.9);
pino.GetYaxis()->SetTitle("DDP (volt)");
pino.Draw();
pina.SaveAs("figura.pdf");
}
```

Code 6.6 Data visualization with ROOT of data acquired in Code 6.5

obtaining the following results (Fig. 6.7).

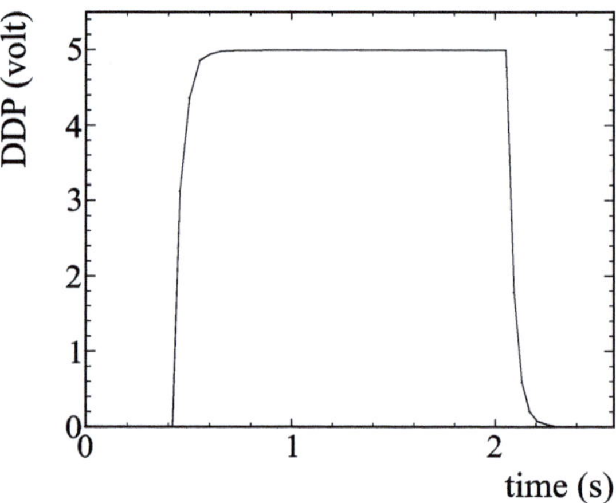

Fig. 6.7 Graphical representation of data in offline mode, after data acquisition

Of course, we can profit from the Python possibilities by using PyROOT, SciPy, matplotlib, and many other libraries to implement the experiment's graphical interface and analyze the data during data acquisition. If we wish to fit the charge and discharge phase of the RC circuit, the following Python script 6.7 shows a possible solution to the problem:

```python
import numpy as np
import matplotlib.pyplot as plt
from numpy import genfromtxt
from scipy.optimize import curve_fit
def func(x, a, b,c):
    return a * (1-np.exp(- (x-c)/b))
def func2(x,a,b,c):
    return a*np.exp(-(x-c)/b)
my_data = genfromtxt('RCSeguenza.dat', delimiter=' '
    )
x = my_data[: , 0]
y = my_data[:, 1]
plt.xlabel('t (s)')
plt.ylabel('V (volt)')
plt.plot(x, y, color = 'black')#charge the first
    plot

x = my_data[(my_data[:,0]<0.74) , 0]
y = my_data[(my_data[:,0]<0.74), 1]
suggest =[5,0.03,0.421]
popt, pcov = curve_fit(func, x, y, p0=suggest)
plt.plot(x, func(x, *popt), 'r-',linestyle='--',
    color = 'b',label='fit: a=%5.3f, b=%7.5f, c=%5.3f
    ' % tuple(popt))

x = my_data[(my_data[:,0]>2.04) , 0]
y = my_data[(my_data[:,0]>2.04), 1]
suggest =[5,0.1,2]
```

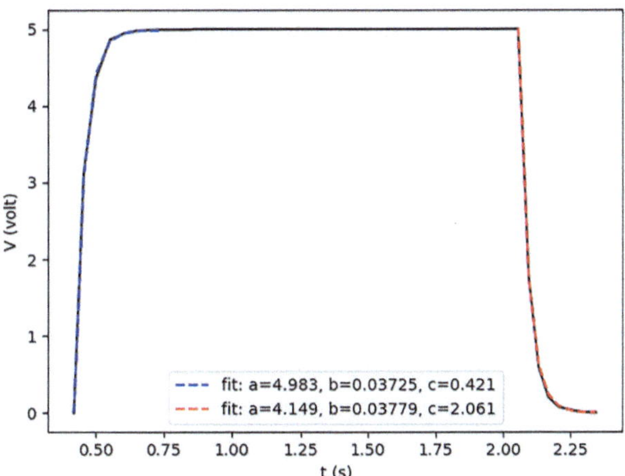

Fig. 6.8 The charge and discharge cycles of an RC circuit as a function of time


```
25  popt, pcov = curve_fit(func2, x, y, p0=suggest)
26  plt.plot(x, func2(x, *popt), 'r-',linestyle='--',
        color='r',label='fit: a=%5.3f, b=%7.5f, c=%5.3f'
        % tuple(popt))
27  plt.legend()
28  plt.show()
```

Code 6.7 Fit data with scipy

The code generates a plot representing the charge and discharge cycles of an RC circuit as a function of time. In this process, the charging and discharging phases are separately modeled using the following mathematical equations:

$$charge(x) = A \cdot (1 - exp(-(x - c)/b)) \tag{6.1}$$

$$discharge(x) = A \cdot exp(-(x - c)/b) \tag{6.2}$$

The charging process follows an exponential function. The term A represents the maximum voltage (or charge) reached after a sufficiently long time, when the capacitor reaches saturation. The parameter b is crucial, as it represents the characteristic time constant τ of the RC circuit, which measures how quickly the circuit charges. The parameter c represents the time shift, accounting for any initial time offset. The code uses these models to fit the charge and discharge process, furnishing the parameters as reported in Fig. 6.8. The values of the b parameter in the two fits present very close values, as expected.

The same results can be obtained with a root macro 6.8 as follows:

```
1      {
2       TCanvas box;
3     TGraph rcplot("RCSeguenza.dat");
4     rcplot.GetXaxis()->SetNdivisions(905);
5     rcplot.GetXaxis()->SetTitle("time (s)");
6     rcplot.Draw();
```

```
 7   rcplot.GetYaxis()->SetTitleOffset(0.9);
 8   rcplot.GetYaxis()->SetTitle("DDP (volt)");
 9   rcplot.Draw();
10   TF1 *charge = new TF1("charge", "[0]*(1-exp(-(x
        -[2])/[1]))",0,0.74);
11   charge->SetParameters(5,0.03,0.421);
12   charge->SetLineColor(4);
13   charge->SetLineStyle(2);
14   TF1 *discharge = new TF1("discharge", "[0]*exp(-(x
        -[2])/[1])",2.045,3);
15   discharge->SetParameters(5,0.1,2);
16   discharge->SetLineColor(2);
17   discharge->SetLineStyle(2);
18   rcplot.Fit("charge","R+");
19   rcplot.Fit("discharge","R+");
20   box.SaveAs("fitrcroot.pdf");
21  }
```

Code 6.8 As in Code 6.7 but with ROOT

6.4.4 Geiger Counter

A Geiger counter spots ionizing radiation with the help of a gas-filled detector. At its
heart sits a Geiger-Müller tube, which essentially behaves like a cylindrical capacitor:
the tube wall is one "plate," a thin axial wire is the other, and the space between them
is filled with low-pressure gas held at roughly 400 V. When a charged particle or
high-energy photon zips through, it knocks electrons off the gas molecules. That tiny
burst of ions triggers a momentary avalanche, creating a sharp current pulse between
the wire and the tube wall. The counter's electronics sense the pulse, record the hit,
and almost instantly recharge the tube so it's ready for the next particle.

This exercise needs a Geiger board like the one presented in Fig. 6.9.

This Geiger board in the region labeled with P3 furnishes three pins: the ground,
the 5-Volt supply, and the pin Vin, which was given a TTL negative signal when
a radiation event was registered. This board is easily interfaced with Arduino Uno,
which can provide the power supply and read and count the radiation detection via
the Vin pin. The radiation-induced discharge in the Geiger tube produces a narrow
signal in the order of nanoseconds in time. The electronics of the Geiger counter board
provide a negative TTL signal of approximately 300 microseconds to be detectable
by the Arduino shield.

Like in the previous case, it is possible to implement an Arduino sketch and
visualize the results using a serial monitor and a serial plot. Figure 6.10 reports the
picture of a test with the two boards performed in the laboratory. Like in the previous
case, it is possible to implement an Arduino sketch and visualize the results using
the serial monitor and plot. But it is more convenient to use than before the Firmata

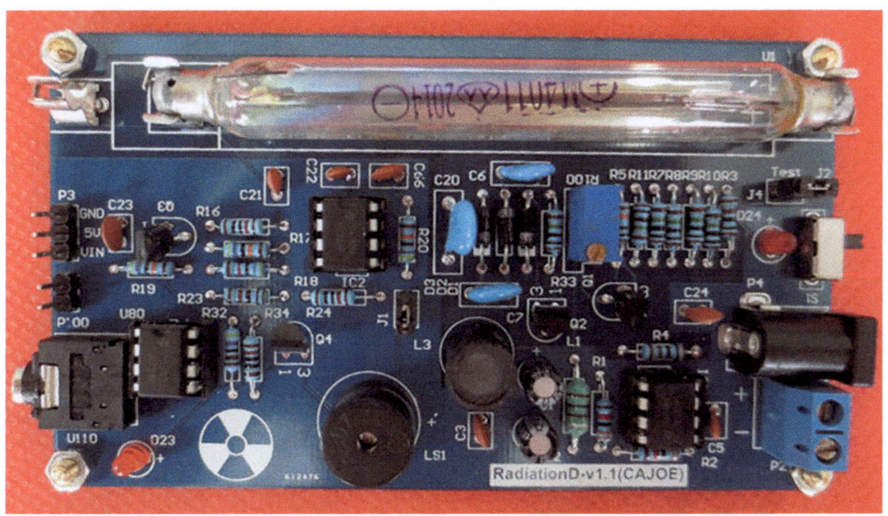

Fig. 6.9 Geiger counter module compatible with Arduino and other microcontroller or microcomputer platforms

firmware was loaded in the Arduino board, and it is easier to use the pyfirmata library to drive the board inside a Python program, running on the PC linked with the board via a USB cable.

The Python code in Listing 6.9 demonstrates the use of the pyfirmata, Numpy, and PyROOT libraries to create a data acquisition interface for a Geiger counter:

- Rows 1–4: Import the necessary libraries.
- Rows 6–22: Set up graphical configurations for the interface using PyROOT.
- Rows 23–26: Define variables and establish connections with the hardware boards.
- Row 34: Begin the polling loop, which continuously checks with the Arduino to detect radiation events.
- Rows 49–54: Implement a visualization of the data acquisition process during the experiment.
- Row 61: Save the collected data to an ASCIIfile using NumPy.
- Rows 62–65: Write the filled histograms to a ROOT file for offline analysis.

Fig. 6.10 Picture of a Geiger counter controlled by Arduino

```python
import ROOT
import pyfirmata
import time
import numpy as np

h = ROOT.TH1F("myHist", "", 20000, 0, 20)
h2 = ROOT.TH1F("myHist2", "", 20, 0, 1200)
h.GetXaxis().SetTitle("#Delta T (second)")
h.GetXaxis().CenterTitle(1)
h.GetYaxis().SetTitle("Counts")
h.GetXaxis().SetTitleOffset(0.95);
h.GetYaxis().CenterTitle(1)
h.GetYaxis().SetTitleOffset(0.90);
h2.GetXaxis().SetTitle("Time (second)")
h2.GetXaxis().CenterTitle(1)
h2.GetYaxis().SetTitle("#muSv/hr")
h2.GetXaxis().SetTitleOffset(0.95);
h2.GetYaxis().SetTitleOffset(1.05);
h2.GetYaxis().CenterTitle(1)
c1 = ROOT.TCanvas("c1","",600,500)
c2 = ROOT.TCanvas("c2","",600,500)
c2.SetLeftMargin(0.16);
board = pyfirmata.Arduino('/dev/ttyACM0')
it = pyfirmata.util.Iterator(board)
it.start()
digital_input = board.get_pin('d:2:i')
```

```python
previous = time.perf_counter()
tempo = 0.
deltat = 0.
T = []
D = []
contami = 0
#while contami < 10: #condizionato al numero di
    eventi
while tempo < 1200:#condizionato al tempo

    pippo = digital_input.read()
   # print(pippo)
    if pippo is False:
        print(time.perf_counter())
        contami = contami + 1
        now = time.perf_counter()
        deltat = now - previous
        previous = now
        tempo = tempo + deltat
        T.append(tempo)
        D.append(deltat)
        print(tempo,"    ",deltat)
        h2.Fill(tempo,0.0057)#0.0057 peso muSv/hr
        c2.Clear()
        h2.Draw()
        c2.Update()
        h.Fill(deltat)
        c1.Clear()
        h.Draw()
        c1.Update()
        print(time.perf_counter())
        time.sleep(0.001)
npT = np.array(T)
npD = np.array(D)
outputFile = "datiGeiger.dat"
np.savetxt(outputFile, np.column_stack((npT, npD)),
    fmt='%7.4f    %7.4f')
F = ROOT.TFile("geigerambiente_2.root", "recreate")
h.Write()
h2.Write()
F.Close()
```

Code 6.9 Python data acquisition program for Geiger counter

The Code 6.9 computes the histogram of the time difference between a detected event and the subsequent reported in Fig. 6.12, and the measured activity in $\mu Sv/hr$ averaged every 60 s of data acquisition, reported in Fig. 6.11. Figure 6.12 illustrates the exponential trend in the temporal distance between two events, a pattern characteristic of a Poisson process that governs the statistical behavior of rare events. Such events are more likely to occur in close succession. This explains why one often hears clusters of beeps followed by relatively long periods of silence when listening to the characteristic "bip" of a Geiger counter in action. Figure 6.11 shows the average environmental activity over time. One potential application of this exper-

Fig. 6.11 The measured activity in $\mu Sv/hr$ averaged every 60 s of data acquisition

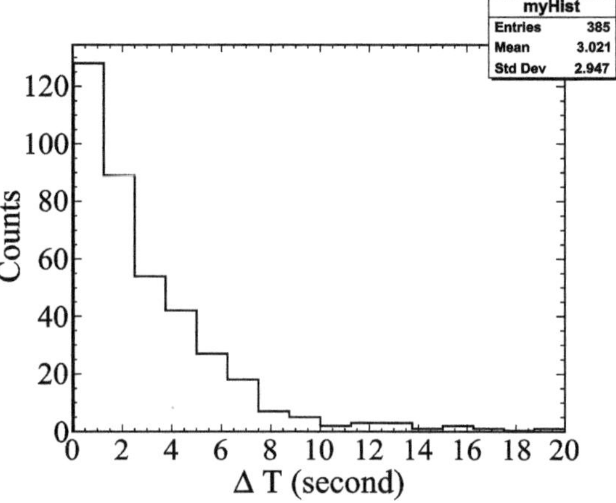

Fig. 6.12 The time difference between a detected event and the subsequent

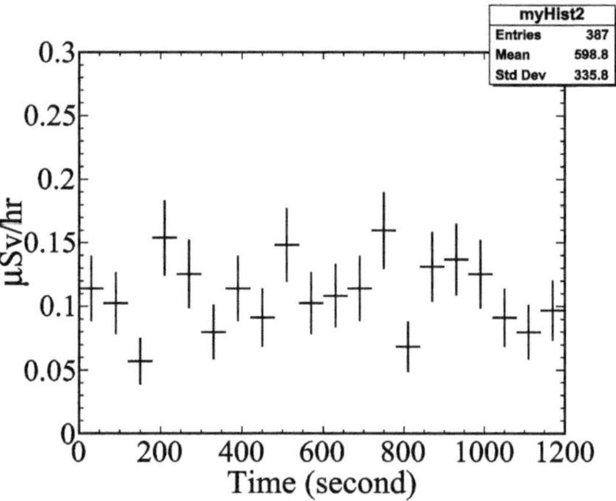

iment and the corresponding plot is radioactivity monitoring. By incorporating a control algorithm that detects deviations from a stable trend (flat line) in the plot, an alarm could be triggered when unexpected variations occur, prompting further investigation into the source of the change. Integrating a power supply, a rechargeable battery, and a photovoltaic panel into the experimental setup makes establishing a self-sustaining, permanent monitoring station possible. This configuration makes the monitoring station reliable and autonomous, ideal for long-term data collection in environmental monitoring or similar applications. The book "Educational and Amateur Geiger Counter Experiments" by Franco Riggi, edited by Springer, contains various interesting experiments with a Geiger counter.

6.5 Exercises

6.1 Perform the same experiment reported in Sect. 6.4.1.1 using Arduino. Measure the temperature of a room every 10 min for a week and then analyze the data. Imagine that the most significant information can be found in the data sample.

6.2 Experiment with the charge and discharge of an RC circuit using Python's "serial" library to make the data-acquisition interface. Collect data and analyze it.

6.3 Mount two Geiger counters in a stack with the two Geiger tubes in the same xy position, collect the output to Arduino's digital ports 2 and 3, and write the code registering only the coincidence between the two detectors.

6.4 Build a meteorological data station (temperature, humidity, pressure) with Raspberry Pi and Arduino.

Solutions

Problems of Chapter 2

2.1

C++:

```cpp
#include <iostream>
using namespace std:
int main(){
    cout<<"I'm writing a message on the screen"<<
    endl;
    return 0;
}
```

Code 6.10 C++ Solution: Write a code displaying a message on the screen

Python:

```python
print("I'm writing a message on the screen")
}
```

Code 6.11 Python Solution: Write a code displaying a message on the screen

2.2

C++:

```cpp
#include <iostream>
using namespace std:
int main(){
    cout<<"Please, give me a number"<<endl;
    float number;
    cin>>number;
    cout<<"the received number is "<<number<<endl;
    return 0;
}
```

Code 6.12 C++ Solution: Write code capable of asking a user for a number and printing it on the screen

© The Editor(s) (if applicable) and The Author(s), under exclusive license to Springer 303
Nature Switzerland AG 2026
S. Vasi et al., *Open Source Tools for Physics Data Analysis*, Undergraduate Texts
in Physics, https://doi.org/10.1007/978-3-032-00721-6

Python:

```python
number = float(input("Please, give me a number"))
print("the received number is ", number)
```

Code 6.13 Python Solution: Write code capable of asking a user for a number and printing it on the screen

2.3

C++:

```cpp
#include <iostream>
using namespace std:
int main(){
    cout<<"Hi! I'm a program summing two numbers"<<
    endl;
    cout<<"Give me the two numebrs"<<endl;
    float number1, number2;
    cin>>number1>>number2;
    cout<<"the the sum of the two numbers is "<<
    number1 + number2<<endl;
    return 0;
}
```

Code 6.14 C++ Solution: Write code capable of asking a user for two real numbers and printing their sum on the screen

Python:

```python
print("Hi! I'm a program summing two numbers")
numbers = list(map(float,input("Give me the two
    numbers").split()))
print("the received number is ", sum(numbers))
```

Code 6.15 Python Solution: Write code capable of asking a user for two real numbers and printing their sum on the screen

2.4

C++:

```cpp
#include <iostream>
using namespace std;
int main(){
    cout<<"HI! I'm a calculator program."<<endl;
    int pass = 1;
    do{
    cout<<"Give me the two numbers"<<endl;
    float number1, number2;
    cin>>number1>>number2;
    cout<<"if you wish to perform a sum digit 1, a
    difference digit 2, a product 3, a ratio 4"<<endl
    ;
    int choice;
    cin >>choice;
    if(choice ==1) {cout<<"the sum of the values is
    "<<number1 +number2<<endl; }
    if(choice ==2) {cout<<"the difference of the
    values is "<<number1 - number2<<endl; }
```

```
15    if(choice ==3) {cout<<"the product of the values
      is "<<number1 +number2<<endl; }
16    if(choice ==4) {cout<<"the ratio of the values
      is "<<number1 +number2<<endl; }
17    cout<<"if you wish another ride digit 1, if not
      whatever numbers"<<endl;
18    cin>>pass;
19    }while(pass==1);
20    return 0;
21  }
```

Code 6.16 C++ Solution: Write code capable of asking a user for two real numbers and asking which of the 4 basic operations they want to perform, then printing the result on the screen. Make the code iterative, capable of repeating the operations unless the user specifies otherwise

Python:

```python
1  print("HI! I'm a calculator program.")
2  while True:#infinite loop
3      numbers = list(map(float,input("Give me the two
       numbers\n").split()))
4      choice = input("if you wish to perform a sum
       digit sum, a difference digit dif, a product pro,
       a ratio rat\n")
5      if choice == "sum":
6          print("the sum of the values is ", sum(
       numbers))
7      if choice == "dif":
8          print("the difference of the values is ",
       numbers[0]-numbers[1])
9      if choice == "pro":
10          print("the product of the values is ",
       numbers[0]*numbers[1])
11      if choice == "rat":
12          print("the ratio of the values is ", numbers
       [0]/numbers[1])
13      if input("if you wish another ride digit yes, if
        not digit whatever\n") != "yes":
14          break
```

Code 6.17 Python Solution: Write code capable of asking a user for two real numbers and asking which of the 4 basic operations they want to perform, then printing the result on the screen. Make the code iterative, capable of repeating the operations unless the user specifies otherwise

2.5
C++:

```cpp
1  #include <iostream>
2  using namespace std;
3  int main(){
4          int number;
5          cout<<"Give me a number and I tell you if is
       odd or even\n";
6          cin>>number;
7    if(number%2==0){
8                  cout<<number<<" is even\n";
```

```cpp
9          }
10         else{
11                 cout<<number<<" is odd\n";
12         }
13  return 0;
14  }
```

Code 6.18 C++ Solution: Write code capable of asking for an integer and indicating whether it is even or odd

Python:

```python
number = int(input("Give me a number and I tell you
    if is odd or even: "))
if (number % 2) == 0:
    print(number," is even")
else:
    print(number," is odd")
```

Code 6.19 Python Solution: Write code capable of asking for an integer and indicating whether it is even or odd

2.6
C++:

```cpp
#include <iostream>
using namespace std;
int main(){
  float data, sum =0;
  cout<<"please, give me 10 numbers and I'give you
     the sum\n";
  for(int i=0; i<10; i++){
   cin>>data;
   sum+=data;
  }
  cout<<"the sum of data is "<<sum<<endl;
  return 0;
  }
```

Code 6.20 C++ Solution:Ask a user to provide 10 numbers and store them in an array. Calculate the sum of the provided numbers and print the final result on the screen

Python:

```python
data = list(map(float,input("please, give me 10
    numbers and I'give you the sum").split()))
print("the sum of data is ", sum(data))
```

Code 6.21 Python Solution: Ask a user to provide 10 numbers and store them in an array. Calculate the sum of the provided numbers and print the final result on the screen

2.7
C++:

```cpp
#include <iostream>
using namespace std;
int main(){
  int solution[3][100];
```

```cpp
  for(int i=0; i<50; i++){
     solution[0][i]=2*i;
     solution[1][i]=2*i+1;
     solution[2][i]=solution[0][i]+solution[1][i];
  }
  for(int i=0; i<50; i++){
     for(int j=0; j<3; j++){
        cout<<solution[j][i]<<"   ";
     }
     cout<<endl;
  }
  return 0;
}
```

Code 6.22 C++ Solution: Write a program filling the first column of a matrix with even numbers from 0 to 100, the second column with odd numbers, and the third column with the sum of the corresponding values from the first two columns

Python:

```python
#by coluns
even = [a for a in range(0,100,2)]
odd  = [a+1 for a in range(0,100,2)]
sum  = [even[a]+odd[a] for a in range(50)]
solution = [even,odd,sum]
print("by columns**********")
print(solution)
#by rows
print("by rows**********")
solution = [[even[a],odd[a],sum[a]] for a in range(50)]
print(solution)
#in C++-Style
print("by rows in c++Style**********")
rows, cols = (50, 3)
solution = [[0 for i in range(cols)] for j in range(rows)]
for i in range(50):
    solution[i][0] = 2*i
    solution[i][1] = solution[i][0] +1
    solution[i][2] = solution[i][0] + solution[i][1]
print(solution)
```

Code 6.23 Python Solution: Write a program filling the first column of a matrix with even numbers from 0 to 100, the second column with odd numbers, and the third column with the sum of the corresponding values from the first two columns

2.8
C++:

```cpp
#include <iostream>
#include <fstream>
#include <vector>
#include <cmath>
using namespace std;
int main(){
  vector<float> data;
  float datum;
  ifstream reader("data.dat");//file in the same
    folder of the code
  if(!reader.is_open()){
    cout<<"I cannot find the file, please check!"<<
    endl;
    return 1;
  }
  while(reader>>datum){
    data.push_back(datum);
  }
  float sum =0;
  for (int i =0; i<data.size(); i++){
    sum+=data[i];
  }
  float mean = sum/data.size();
    for (int i =0; i<data.size(); i++){
    sum+=data[i];
  }
    sum =0;
    for (int i =0; i<data.size(); i++){
    sum+=pow(data[i]-mean,2);
  }
  float std = sqrt(sum/data.size());
  cout<<"the mean value is "<<mean<<" the standard
    deviation "<<std<<endl;
  return 0;
  }
```

Code 6.24 C++ Solution: Write code that reads the file "data.dat," prints all the values on the screen, and calculates the mean and standard deviation

Python:

```python
import math
data = list(map(float,open("data.dat").read().split
    ()))
mean = sum(data)/len(data)
gaps = [(data[i]-mean)**2. for i in range(len(data))
    ]
rms = math.sqrt(sum(gaps)/len(gaps))
print("the mean value is ",mean," the standard
    deviation ",rms)
```

Code 6.25 Python Solution: Write code that reads the file "data.dat," prints all the values on the screen, and calculates the mean and standard deviation

2.9

C++:

```cpp
#include <iostream>
#include <fstream>
#include <sstream>
#include <string>
using namespace std;
int main(){
  string line, number;
  ifstream reader("data.dat");//file in the same
    folder of the code
  if(!reader.is_open()){
    cout<<"I cannot find the file, please check!"<<
    endl;
    return 1;
  }
  getline(reader,line);
  stringstream stringer(line);
  int columns=0;
  while(stringer>>number){
    columns++;
  }
    int rows = 1;
    while(getline(reader,line)){
    rows++; //mi conto tutte le righe
    }
    cout<<"number of rows is "<<rows<<" numbers of
    columns is "<<columns<<endl;
  return 0;
  }
```

Code 6.26 C++ Solution: Write code that counts the rows and columns in the file "data.dat"

Python:

```python
data =open("data.dat").read()
rowfinder = data.split("\n")
rows = len(rowfinder)-1 #remove the last new line
columns = len(rowfinder[0].split())
print("number of rows is ",rows," numbers of columns
    is ",columns)
```

Code 6.27 Python Solution: Write code that counts the rows and columns in the file "data.dat"

2.10

C++:

```cpp
#include <iostream>
#include <fstream>

using namespace std;
int main(){

  ifstream reader("data.dat");//file in the same
    folder of the code
```

```cpp
 8    if(!reader.is_open()){
 9      cout<<"I cannot find the file, please check!"<<
      endl;
10      return 1;
11    }
12
13    return 0;
14  }
```

Code 6.28 C++ Solution: Write code finding the absolute maximum and minimum of the fifth column of the file "data.dat"

Python:

```python
1 data1 =open("data.dat").read().split("\n")
2 data2 = [list(map(float,data1[i].split())) for i in
      range(len(data1)-1)]
3 data = []
4 for a in range(len(data2)):
5     data.append(data2[a][1])
6 print(data)
7 data.sort()
8 print(data[0])
9 print(data[len(data)-1])
```

Code 6.29 Python Solution: Write code finding the absolute maximum and minimum of the fifth column of the file "data.dat"

3.1

```cpp
 1 #include <iostream>
 2 using namespace std;
 3 class changer{
 4 float rateEtoD=1.08 ;
 5 float rateEtoF=0.95 ;
 6        public:
 7        changer(){
 8                cout<<"rate Euro to US-Dollar I know
    is "<<rateEtoD<<endl;
 9                cout<<"rate Euro to Swiss-Franc I
    know is "<<rateEtoF<<endl;
10        }
11        changer(float a, float b){
12                rateEtoD=a;
13                rateEtoF=b;
14        }
15        float ConvEtoD(float a){
16                return a*rateEtoD;
17        }
18        float ConvDtoE(float a){
19                return a/rateEtoD;
20        }
21        float ConvEtoF(float a){
22                return a*rateEtoF;
23        }
24        float ConvFtoE(float a){
```

```
25                    return a/rateEtoF;
26            }
27            float ConvFtoD(float a){
28                    return ConvEtoD(ConvFtoE(a));
29            }
30            float ConvDtoF(float a){
31                    return ConvEtoF(ConvDtoE(a));
32            }
33
34 };
35 int main(){
36            changer pino;
37            float euro;
38            cout<<"how many euros you wish to change?\n"
       ;
39            cin>>euro;
40            cout<<"corresponds in US-dollar to "<<pino.
       ConvEtoD(euro)<<endl;
41            cout<<"corresponds in Swiss-Franc to "<<pino
       .ConvEtoF(euro)<<endl;
42
43 return 0;
44 }
45 ~
```

Code 6.30 Root solution: Write a class of handling currency conversion between the euro, dollar, and Swiss franc. The class must contain methods for loading data in any currency, returning data in any currency, and the variables storing the current exchange rates for conversions. The class must be tested in a working program with an example that proves its functionality

5.1
ROOT:

```
1  Double_t funz(Double_t *x, Double_t *par){
2            if (x[0]<0) return sin(x[0])/x[0];
3            if(x[0] >0) return -sin(x[0])/x[0];
4            return 0;
5
6  }
7  void fun()
8  {
9  TCanvas *box= new TCanvas("box","",600,500);
10 TF1 *myfun= new TF1("myfun",funz,-10,10);
11 myfun->Draw();
12
13 return;
14 }
```

Code 6.31 Root solution: Plotting a function defined in different intervals

Python:

```python
import matplotlib.pyplot as plt
import numpy as np
x1 = np.linspace(-10, 0.0001,100)
x2 = np.linspace(0.0001, 10,100)
x =x1 + x2
y1 = np.sin(x1) / x1
y2 = -np.sin(x2) / x2
y=y1+y2
plt.plot(x,y)
plt.show()
```

Code 6.32 Python solution: Plotting a function defined in different intervals

Index

A
Algorithm, 46, 105, 106, 143, 146, 152, 161,
178, 180, 182, 199, 262, 266, 280, 300

Append, 167, 168
Arduino, 278–281, 283, 287, 288, 291–293,
296, 297
firmata, 292
sketch, 296, 297
ASCII, 27, 32, 94, 132, 134, 297
Assignment, 23, 45, 46
Attributes, 75, 76, 90, 165, 209, 252
Axis, 125, 138, 210–212, 214, 216, 217, 221,
223, 224, 226, 241, 242, 249, 251, 254,
259, 274

B
Binary, 25, 29, 94, 100, 131, 134, 162, 166,
281
Boolean, 22, 24, 28, 50, 139, 207
Boot, 8, 9

C
C++, 6, 19, 20, 22, 28, 53–55, 57, 58, 61,
76, 78, 80, 81, 89–91, 94, 99–101, 115,
119, 121, 267, 279–281, 283, 287, 288,
292–294
Array, 36
Base types, 22
Compiler, 20

Control structures, 46
Dictionary, 43
Editor, 20
Functions, 65, 67, 69
Histogram, 63
I/O operations, 30
List, 38
Loop, 50
Operators, 44, 45
Class, 21, 23, 28, 32, 33, 35, 36, 38, 55, 57,
61, 75, 76, 78–82, 89–92, 94, 95, 100,
101, 104, 106, 117–121, 123, 126, 128,
131, 132, 134–136, 139, 142, 156, 161,
162, 193, 247, 251, 267, 274
constructors, 78
destructor, 78
in Python, 89, 91, 92
Inheritance, 81
methods, 76, 78
private, 78, 80
public, 80
Command Line, 11, 16, 107, 112, 114, 115,
162, 235, 247, 279
Interface (CLI), 4, 7, 10, 235

E
Enviro+, 288–290
Environment, 280, 281, 283
monitoring, 288
Sensors, 288